Andrew Stephenson

Public Lands and Agrarian Laws of the Roman Republic

Andrew Stephenson

Public Lands and Agrarian Laws of the Roman Republic

ISBN/EAN: 9783337008895

Printed in Europe, USA, Canada, Australia, Japan

Cover: Foto ©Suzi / pixelio.de

More available books at **www.hansebooks.com**

JOHNS HOPKINS UNIVERSITY STUDIES

IN

HISTORICAL AND POLITICAL SCIENCE

HERBERT B. ADAMS, Editor

History is past Politics and Politics present History—*Freeman*

NINTH SERIES

VII-VIII

PUBLIC LANDS AND AGRARIAN LAWS

OF THE

ROMAN REPUBLIC

BY ANDREW STEPHENSON, PH. D.

Professor of History, Wesleyan University

BALTIMORE

THE JOHNS HOPKINS PRESS

JULY-AUGUST, 1891

COPYRIGHT, 1891, BY THE JOHNS HOPKINS PRESS.

JOHN MURPHY & CO., PRINTERS.
BALTIMORE.

PREFACE.

In the following pages it has been my object to trace the history of the domain lands of Rome from the earliest times to the establishment of the Empire. The plan of the work has been to sketch the origin and growth of the idea of private property in land, the expansion of the *ager publicus* by the conquest of neighboring territories, and its absorption by means of sale, by gift to the people, and by the establishment of colonies, until wholly merged in private property. This necessarily involves a history of the agrarian laws, as land distributions were made and colonies established only in accordance with laws previously enacted.

My reason for undertaking such a work as the present is found in the fact that agrarian movements have borne more or less upon every point in Roman constitutional history, and a proper knowledge of the former is necessary to a just interpretation of the latter.

This whole question presents numerous obscurities before which it has been necessary more than once to hesitate; it offers, both in its entirety and in detail, difficulties which I have at least earnestly endeavored to lessen. These obscurities and difficulties, arising in part from insufficiency of historical evidence and in part from the conflicting statements of the old historians, have been recognized by all writers and call forth on my part no claim for indulgence.

This monograph is intended as a chapter merely of a history of the public lands and agrarian laws of Rome, written for the purpose of a future comparison with the more recent agrarian movements in England and America.

ANDREW STEPHENSON.

MIDDLETOWN, CONN.
May 8, 1891.

TABLE OF CONTENTS.

CHAPTER I.
PAGE.
SEC. 1. LANDED PROPERTY.. 7
" 2. QUIRITARIAN OWNERSHIP... 13
" 3. AGER PUBLICUS.. 15
" 4. ROMAN COLONIES... 19

CHAPTER II.
SEC. 5. LEX CASSIA... 24
" 6. AGRARIAN MOVEMENTS BETWEEN 486 AND 367.............. 26
    (a). Extension of Territory of conquest up to the year 367 B. C... 35
    (b). Colonies Founded between 454 and 367................... 36
SEC. 7. LEX LICINIA... 36
" 8. AGRARIAN MOVEMENTS BETWEEN 367 AND 133.............. 46
    (a). Extension of Territory by conquest between 367 and 133.. 59
    (b). Colonies Founded between 367 and 133................... 60
SEC. 9. LATIFUNDIA... 62
" 10. INFLUENCE OF SLAVERY.. 66
" 11. LEX SEMPRONIA TIBERIANA.. 69
" 12. LEX SEMPRONIA GAIANA... 77

CHAPTER III.
SEC. 13. LEX THORIA.. 79
" 14. AGRARIAN MOVEMENTS BETWEEN 111 AND 86.............. 88
" 15. EFFECT OF THE SULLAN REVOLUTION........................ 91
" 16. AGRARIAN MOVEMENTS BETWEEN 86 AND 59................ 93
" 17. LEX JULIA AGRARIA.. 95
" 18. DISTRIBUTIONS OF LAND AFTER THE CIVIL WAR BETWEEN CÆSAR AND POMPEY.. 98
" 19. DISTRIBUTIONS FROM THE DEATH OF CÆSAR TO THE TIME OF AUGUSTUS... 99
    (a). Lex Agraria of Lucius Antonius............................ 99
    (b). Lex de Colonis in Agros Deducendis....................... 99
    (c). Second Triumvirate.. 100

PUBLIC LANDS AND AGRARIAN LAWS OF THE ROMAN REPUBLIC.

CHAPTER I.

Sec. 1.—Landed Property.

The Romans were a people that originally gave their almost exclusive attention to agriculture and stock-raising. The surnames of the most illustrious families, as Piso (miller), Porcius (swine-raiser), Lactucinius (lettuce-raiser), Stolo (a shoot), etc., prove this. To say that a man was a good farmer was, at one time, to bestow upon him the highest praise.[1] This character, joined to the spirit of order and private avarice which in a marked degree distinguished the Romans, has contributed to the development among them of a civil law which is perhaps the most remarkable monument which antiquity has left us. This civil code has become the basis of the law of European peoples, and recommends the civilization of Rome to the veneration of mankind.

The corner-stone of this legislation was the constitution of the law of property.[2] This property applies itself to every-

---

[1] Cato, *De Re Rustica*, I, lines 3–8. "Majores nostri .... virum bonum cum laudabant, ita laudabant, bonum agricolam bonumque colonum. Amplissime laudari existimabatur, qui ita laudabatur."

[2] Muirhead, *Roman Law*, 36 *et seq.*

thing in the law of Rome, to land, to persons and to obligations.

Urbs, the name of the village, takes its origin, according to an etymology given by Varro,[1] from the furrow which the plow traced about the habitations of the earliest dwellers. But what is of more interest to us is that the legal signification of *Urbs* and *Roma* was different. The former was the village comprised within the sacred enclosure; the latter was the total agglomeration of habitations which composed the village, properly[2] so called, and the outskirts, or suburbs. The powers of certain magistrates ceased with the sacred limits of the *Urbs*, while the privileges accorded to a citizen of Rome extended to the village and the suburbs and finally embraced the entire Roman world.

The most ancient documents which have reached us from the history of India and Egypt reveal that they had landed property fully established, while Roman annals reveal to us the very creation of this institution. Whatever modern criticism may deduce, Dionysius, Plutarch, Livy, and Cicero agree in representing the first king of Rome as merely establishing public property in Roman soil. This national property, the people possessed in common and not individually. Such appears to us to be the quiritarian property *par excellence*,[3] and its primitive form was a variety of public community[4] of which individual property was but a later solemn emancipation. To this historic theory attaches the true notion of quiritarian land of which we will speak in greater detail hereafter.

As regards the organization and constitution of individual and private property, the traditions themselves attribute this to the second king of Rome, the real founder of Roman

---

[1] Varro, *De Lingua Latina*, V, 143.
[2] Frag. to Digest, 287 and 147 of Title 16, Bk. 50 with notes of Schultung and Small.
[3] Plutarch's *Romulus*, § 19. [4] Mommsen, *History of Rome*, 1, 194.

society, who divided the territory among the citizens, marking off the limits of individual shares and placing them under the protection of religion. In this way a religious charter was granted to the institutions of private property. Thus a primitive division of territory appears to have been the basis of these varied traditions, but the precise form of this division eludes us.

The Roman territory was confined for many ages to a surface of very limited extent, which properly bore the name of *Ager Romanus*. This name with signification slightly changed appeared to be still in use in the time of the empire, and even at the present day a portion of the Roman territory which very nearly corresponds to the ancient territory of the imperial period is called *Agro Romano*.[1] That which was properly called *Ager Romanus* at first only occupied the surface of a slightly expanded arc whose chord was the river Tiber.[2] Primitive Rome did not extend beyond the Tiber into Etruria, and toward Latium her possessions did not extend beyond the limits of some five or six miles reckoning from the Palatine. Toward the east the towns of Antemnae, Fidenae, Caenina, Collatia and Gabia lay in the immediate neighborhood, thus limiting the extension of the city in that direction within a radius of five or six miles;[3] and northward the Anio[4] formed the limit. To the southwest as you approach Lavinium, the sixth milestone marked the boundary of Rome. Thus with the possible exception of a small strip of land extending upon either bank of the Tiber to its mouth, and embracing the old site[5] of Ostia, have we marked out all of ancient Rome. Strabo[6] says it could be gone round in a single day. And according to this same author it was within these limits that the annual auspices[7] could be taken.

---

[1] Sismondi, *Etudes sur l'econ. polit.*, I, 2, § 1.
[2] Pseudo Fabius Pictor, Bk. I, p. 54; Plut., *Numa*, 16; Festus V° Pectustum Palati, p. 198 and 566, Lindemann.
[3] Arnold, *Roman History*, I, ch. 3, par. 4.   [4] Mommsen, I, 75.
[5] Strabo, Bk. 5, 253.   [6] Strabo, Bk. 5, ch. 3, § 2.   [7] Arnold, I, ch. 3.

Both city and land increased with time. Property seemed to have been added and lost successively during the reign of the kings.[1] The last increase of the *Ager Romanus* was due to the labors of Servius Tullius, and it was in the reign of this king that it reached its greatest limit. Dionysius[2] says: "As soon as he (Servius) was invested with the government, he divided the public lands among such of the Romans as having no lands of their own, cultivated those of others. . . . He added two hills to the city, that called the Viminal and the Esquiline hill, each of which forms a considerable city; these he divided among such Romans as had no houses, to the intent that they might build them. . . . This king was the last who enlarged the circumference of the city by the addition of these two hills to the other five, having first consulted the auspices as the law decided, and performed the other religious rites. Further than this the city has not since then been extended." Without doubt these possessions received great additions in later times,[3] but they were not incorporated in the *Ager Romanus* as the preceding had been. The subjugated territories kept their ancient names while their lands were made the object of distributions to the people, of public sales to the citizens who also extended their possessions outside of Roman[4] territory, or else the new conquests were abandoned to municipia, given up to colonies, or became a part of that which was called *Ager Publicus*. In fine, it was a fundamental principle of the public law of Rome that the lands and the persons of the people conquered belonged to the conqueror, the Roman people, who either in person or by their delegates disposed of them as it seemed best. Among the ancients war always decided concerning both liberty and property.

---

[1] Dionysius, II, 55; V, 33, 36; III, 49-50; Livy, I, 23-36.
[2] Dionysius, IV, 13.
[3] Varro, *De Lingua Latina*, V, 33.
[4] Sigonius, *De Antiq. Juris Civ. Rom.*, Bk. I, ch. 2.

The result of all these facts was that the Roman territory was made the object of a division or a primitive distribution either among the three races of the first population, or a little later among the citizens or inhabitants. This very same principle has been frequently observed in recent times in regard to confiscated [1] territories and conquered peoples.

Now what was the allotment of the first distribution of land?

Upon this topic the ancient authorities are blind and confusing to such an extent as to be wholly inadequate for the solution of the difficulty. Among the more recent authorities, two opposing systems have been sustained, the one represented by Montesquieu, and the other by Niebuhr. (1) According to Montesquieu, the kings of Rome divided the land into perfectly equal lots for all the citizens and the title of the law of the Twelve Tables relative to successions was for no other object than to establish this ancient equality of the division of lands.[2] (2) Niebuhr,[3] on the contrary, claimed that territorial property was primitively the attribute of the patriciate and everyone who was not a member of this noble race was incapable of possessing any part of the territory. From this theory the author deduced numerous consequences which are important both to law and history. Neither of these systems is free from errors. Montesquieu seems to have made no difference between patrician and plebeian in using the term *citizen*, while it is no longer disputed that the plebeian was not a burgess and consequently had no civic rights save those granted to him by the ruling class. His idea of goods must have, at least, become chimerical at a very early date, as this equality was so little suspected by the ancients that Plutarch,[4] after

---

[1] Hume's *Hist. of Eng.*, I, ch. 4; IV, ch. 61.
[2] *Esprit des lois*, Liv. 27, c. 1.
[3] *Roman Hist.*, II, 164; III, 175 and 211.
[4] Lycurgus and Numa, II; Cicero, *De Repub.*, 11, 9.

having spoken of the efforts of Lycurgus to overturn the inequality of wealth among the Spartans, accuses Numa of having neglected a necessity so important. It is moreover difficult to see how Montesquieu could think that testamentary disposition tended to maintain equality when the privilege was accorded to every citizen of disposing of his entire patrimony by will even to the prejudice of his children.[1] Again, the law of debts was hardly favorable[2] to equality.

Niebuhr clearly[3] denied the existence of the plebs until Ancus incorporated the Latins and bestowed upon them peculiar privileges thus forming a new and third class distinct from both patricians and clients. Had Niebuhr succeeded in establishing this view, the right to landed property would appear to be wholly vested in the patricians, for a client, from the very nature of his position, could hold nothing independent of his master. But this theory has fallen to the ground and no writer of the present day pretends to uphold it. The plebeians existed from the very first and some of them held land in full private ownership very little different from the quiritarian ownership of the patricians. Cicero, who in his Republic has occupied himself with the ancient constitution of Rome and has spoken in detail of the division of the lands, always speaks of the distribution among the citizens without regard to quality of patrician or plebeian, *divisit viritim civibus*. He has nowhere written that territorial riches were the exclusive appanage of the patriciate. It must be confessed, however, that it is doubtful whether he intended to embrace the plebeians in his *civibus*. For more than two centuries before the time of Cicero the plebeians had enjoyed the full rights of Roman citizenship, but for more than that length of time property

---

[1] Muirhead, *Roman Law*, 46 and note—"uti legasset suae rei ita jus esto."
[2] Muirhead, 92–96.   [3] Niebuhr, I.
[4] Momm., I, 126; Ihne, I; Nitzsch, *Geschichte der römischen Republik*, 52; Lange, *Römische Geschichte*, I, 18.

had been concentrated in the hands of the aristocracy. This result was the consequence of the Roman constitution[1] and the establishment of a populous city in the midst of a narrow surrounding country. Roman policy had never been conducive to this concentration, and it will hereafter appear that the nobility who had the chief direction and administration of public affairs had little by little usurped the property which formed the domain of the state, *i. e.* *Ager Publicus,* and swallowed up the revenues due the treasury.

SEC. 2.—QUIRITARIAN OWNERSHIP.

Citizenship was the first requisite to the right of property in Roman territory. This rule, although invariable and inherent in the Roman state, bent under the influence of international politics or the philosophy of law, yet its severity affords us a notable characteristic of the law of ancient Rome. Cicero and Gaius have preserved to us an important monument of this law in a fragment of the Twelve Tables which proclaims the solemn principle, *adversus hostem aeterna auctoritas esto.*[2] *Hostis* in the old Latin language was synonymous with stranger, *perigrinus.*[3] This Roman name was moreover applied to a person who had forfeited the protection of the law by reason of a criminal condemnation, and who was therefore designated *peregrinus.*[4]

Auctoritas also had in old Latin a different signification from what it has in later Latin. It expressed the idea of the right to claim and defend in equity. It was very nearly equivalent to the right of property.[5] The sense of the Roman

---

[1] Dureau de la Malle, *Mém. sur les pop. de l' Italie,* 500 *et seq.*
[2] De Officiis, I, 12; Gaius, Frag., 234: Digest, 50, 16.
[3] Varro, De L. L. V. 14; Plautus, *Trinummus,* Act I, Scene 2, V. 75; Harper's *Latin Dictionary;* Cicero, *De Off.*, I, 12: "Hostis enim apud majores nostros is dicibatur, quem nunc peregrinum dicimus."
[4] Cic., *loc. cit.*; Gaius, Frag., 234.
[5] Forcellini, *Lexic.;* Harper's *Latin Lex.*

law was, then, that the *peregrinus* could not bar or proceed against a Roman, a disposition somewhat similar to the old law of England.[1] And as it was necessary to be a citizen in order to acquire by the civil and solemn means which dominated the law of property in Rome, it followed that the *peregrini* were excluded from all right to property in land by these laws. This exclusive legislation for a long time governed Europe and did not disappear even from the Code Napoleon of 1819.[2]

We have a forcible example of the severity of the old Roman law in this regard in the text of Gaius,—*Aut enim ex jure quiritium unusquisque dominus erat, aut non intelligebatur dominus.*[3]

Dominium was therefore inseparable from *Jus Quiritium*, the law of the Roman city, the *optimum jus civium Romanorum*. The *peregrinus* was excluded from landed property both Roman and private; he could neither inherit nor transmit; claim nor defend in equity. Moreover the name *peregrinus* was not confined to the stranger proper but was also bestowed upon subjects of Rome[4] who, being deprived of their property and also of political liberty by right of conquest, had not received the right of citizenship which was for a long time confined within very narrow limits. It would thus appear conclusive from the law quoted that the client and plebeian could not at first hold land *optimo ex jure quiritium*.

Thus the tenure of the patricians was three-fold: first, they had full property in the land; second, they had a seigniorial right, *jus in re*, in the land of their clients and the plebeians whose property belonged to the *populus*, *i. e.* the generality of the patricians; in the third place, in their own hands, they

---

[1] *i. e.* The descendents of a person escheated could bring no action for the recovery of the property.
[2] Giraud, *Recherches sur le Droit de Propriété*, p. 210.
[3] Gaius, Bk. II, 40.
[4] Ulpian, Frag., Title XIX, 4; Giraud, 216.

held lands which were portions of the domain and which were held by a very precarious tenure called *possessio*.

According to Ihne, all lands in Rome were held by the above mentioned tenure until the enactment of the Icilian law *de Aventino publicando* which involved a change of tenure by converting the former dependent and incumbered tenure of the plebeians into full property.

SEC. 3.—AGER PUBLICUS.

In her early history Rome was continually making fresh conquests, and in this way adding to her territory.[1] She steadfastly pursued a course of destruction to her neighbors in order that she might thereby grow rich and powerful. In this way large tracts of territory became Roman land, the property of the state or *Ager Publicus*.[2]

This public land extended in proportion to the success of the Roman arms, since the confiscation of the territory of the vanquished was, in the absence of more favorable terms, a part of the law of war. All conquered lands before being granted or sold to private individuals were *Ager Publicus*,[3] a term which with few exceptions came to embrace the whole Roman world.

This *Ager Publicus* was farther increased by towns[4] voluntarily surrendering themselves to Rome without awaiting the iron hand of war. These were commonly mulcted of one-third of their land.[5] "The soil of the country is not the product of labor any more than is water or air. Individual citizens cannot therefore lay any claim to lawful property in land as to anything[6] produced by their own hands." The state in this case, as the representative of the rights and

---

[1] Long, *Decline of the Roman Rep.*, I, ch. 11.
[2] Muirhead, *Roman Law*, 92.
[3] Ortolan, *Histoire de la legislation Romaine*, p. 21.
[4] Mommsen, I, 131; Arnold, I, 157.
[5] Dionysius, IV, 11, Livy.   [6] Ihne, I, 175.

interests of society, decides how the land shall be divided among the members of the community, and the rules laid down by the state to regulate this matter are of the first and highest importance in determining the civil condition of the country and the prosperity of the people. Whenever but one class among the people is privileged to have property in land a most exclusive oligarchy is formed.[1] When the land is held in small portions by a great number and nobody is legally or practically excluded from acquiring land, there we find provided the elements of democracy.

According to the strictest right of conquest in antiquity the defeated lost not only their personal freedom, their moveable and landed[2] property, but even life itself. All was at the mercy of the conquerors. In practice a modification of this right took place and in Rome extreme severity was applied only in extreme cases, generally as a punishment for treason.[3]

This magnanimity was not rare and it even went so far as to restore the whole of the territory to the people subdued.[4] But let us not suppose that this humanity toward a conquered people sprang from any pity inspired by their forlorn condition. It was due merely to the interest of the conquerors themselves. The conquered lands must still be cultivated and the depleted population restored. For this reason the conquered had generally not only life and freedom left them but also the means of livelihood, i. e. some portion of their land. This portion they held subject to no restrictions or services save those levied upon quiritarian property. It was private property to the full legal extent of the expression, thus being in the unlimited disposition of the individual.[5] These people formed the nucleus of the plebeians, the freemen who were

---

[1] Ihne, I, 175.
[2] Livy, Bk. I, c. 38, with note by Drachenborch; Livy, Bk. VII, c. 31.
[3] Siculus Flaccus, *De Conditione Agrorum*, 2, 3: "Ut vero Romani omnium gentium potiti sunt, agros alios ex hoste captos in victorem populum partiti sunt, alios verro agros vendiderunt, ut Sabinorum ager qui dicitur quaestorius." [4] Cicero, in Verrem, II, Bk. 3, § 6.
[5] Giraud, *Droit de propriété chez les romains*, 160.

members of the Roman state[1] without actually having any political rights.

The *Ager Publicus* was the property of the state and as such could be alienated only by the state.[2] This alienation could be accomplished in two ways:

(*a*). By public sale;
(*b*). By gratuitous distribution.

(*a*). The public sale was merely an auction to the highest bidder and in the later days of the monarchy and early part of the republic, rich plebeians must have become possessed of large tracts of land in this way; the privilege of acquiring property in land having been extended to them some time before the Servian reform.[3]

(*b*). The gratuitous distribution of land was accomplished by means of Agrarian Laws or royal grant and had for its object the establishment of colonies for purposes of defence, the rewarding of veterans or meritorious soldiers,[4] or in later times, the providing for impoverished plebeians.

But even in the earliest times a portion of the domain lands was excluded from sale or private appropriation,[5] in order to serve as a resource for the needs of the state.

This was the general usage of ancient republics and this maxim of reserved lands was recommended[6] by Aristotle as the first principle of political economy.

Such reserved *ager publicus* was leased either in periods of five years (quinquennial leaseholds) or perpetually, *i. e.*, by emphyteutic lease or copyhold. From these lands[7] the treasury received an income of from one-tenth to one-fifth of the annual crops.

Besides these legal methods mentioned there was another very common one which was seemingly never established by any law and therefore existed merely by title of tolerance. I

---

[1] Ihne, I, 175. [2] Muirhead, 92; Giraud, 165.
[3] Higin., *De Limit. Const. apud Goes. Rei Agr. Script.*, pp. 159–160.
[4] Giraud, 164. [5] Dionysius, II, 7.
[6] Aristotle, *Polit.*, Z. Κεφ. θ. 7: Αναγκαιὸν τοίνυν εἰς δυο μέρη διηρῆσθαι τὴν χώραν καὶ τὴν μὲν εἶναι κοινὴν, τὴν δὲ τῶν ἰδιωτῶν. [7] Giraud, 163.

speak of the indefinite *possessio* which was nothing but an occupation on the part of the patricians[1] of the land belonging to the state and was in nature quite similar to the so-called "squatting" commonly practiced in some of our western states and territories. The title to the enjoyment of the public lands was at first clearly vested in the patricians nor was this right extended to the plebeians until after they had been admitted to full citizenship. With regard to the state the *possessor*[2] was merely a tenant at will and could be removed whenever desired; but as regarded other persons he was like the owner of the soil and could alienate the land which he occupied either for a term of years, or forever, as if he were the real proprietor.[3] The public land thus occupied was looked to as a resource upon the admission of new citizens. They customarily received a small freehold according to the general notion of antiquity that a burgess must be a landowner. This land could only be found by a divison of that which belonged to the public, and a consequent ejectment of the tenants at will. In the Greek states every large accession to the number of citizens was followed by a call for a division of the public lands and, as this division involved the sacrifice of many existing interests, it was regarded with aversion by the old burgesses as an act of revolution.

A great part of the wealth of the Romans consisted in domains of this kind, and the question will occur to the thoughtful mind how the government was able to keep the most distinguished part of her citizens in a legal position so uncertain and alarming. English law is very different from the Roman in this respect and would decide in favor of the

---

[1] Festus, p. 209, Lindemann; Cicero, ad Att. II, 15; Philipp. V, 7; De Leg. Agr. I, 2, III, 3; De Off. II, 22; Livy, II, 61, IV, 51, 53, VI, 4, 15; Suet. Julius Cæsar, 38; Octavius, 13, 32; Cæsar, De Bell. Civ., I, 17; Orosius, V, 18.

[2] Aggenus Urbicus, p. 69, ed. Goes.

[3] Giraud, 185–187; Mommsen, I, 110; Ortolan, 227; Hunter, *Roman Law*, 367.

tenant and against the state. It is fairly possible that this uncertainty of tenure tended to render the government more stable and less liable to sudden revolutionary movements, thus having the same effect upon the Roman government which funded debts have upon the nations of to-day.

SEC. 4.—ROMAN COLONIES.

Probably in no other way does the Roman government so clearly reveal its nature and strength as in its method of colonization. No other nation, ancient or modern, has ever so completely controlled her colonies as did the Roman. Her civil law, indeed, reflected itself in both political and international relations. In Greece, as soon [1] as a boy had attained a certain age his name was inscribed upon the tribal rolls and henceforth he was free from the *potestas* of his father and owed him only the marks of respect which nature demanded. So too, at a certain age, the colonies separated themselves from their mother city without losing their remembrance of a common origin. This was not so in Rome. The children [2] were always under the *potestas* of their parents. By analogy therefore, the colonies ought to remain subject to their mother city. Greek colonies went forth into a strange land which had never been conquered by Hellenic arms or hitherto trod by Grecian foot. Roman [3] colonies were established by government upon land which had been previously conquered and which therefore belonged to the Roman domain. The Greek was fired with an ambition to obtain wealth and personal distinction, being wholly free to bend his efforts to personal ends. Not so the Roman. He sacrificed self for the good of the state. Instead of the allurements of wealth he received some six jugera of land, free from taxation it is true, but barely

---

[1] Bouchaud, M. A., *Dissertation sur les colonies romaines*, pp. 114–222, en Memoires de l'institut Sciences, Morals et Politique, III.
[2] Muirhead's Article on *Roman Law* in Ency. Brit.; Ihne, I, 235.
[3] Momm., I, 145.

enough to reward the hardest labor with scanty subsistence. Instead of the hope of personal distinction, he in most cases sacrificed the most valuable of his rights, *jus suffragii et jus*[1] *honorum* and suffered what was called *capitis diminutio*. He devoted himself, together with wife and family, to a life-long military service. In fact the Romans used colonization as a means to strengthen their hold upon[2] their conquests in Italy and to extend their dominion from one centre over a large extent of country. Roman colonies were not commercial. In this respect they differed from those of the Phoenicians and Greeks. Their object was essentially military[3] and from this point of view they differed from the colonies of both the ancients and moderns. Their object was the establishment of Roman power. The colonists marched out as a garrison into a conquered town and were exposed to dangers on all sides. Every colony acted as a fortress to protect the boundary and keep subjects to their allegiance to Rome. This establishment was not a matter of individual choice nor was it left to any freak of chance. A decree of the senate decided when and where a colony should be sent out, and the people in their assemblies elected individual members for colonization.

From another point of view Roman colonies were similar to those of Greece, since their result was to remove from the centre to distant places the superabundant population, the dangerous,[4] unquiet, and turbulent.

But the difference in the location of the colonies was easy to distinguish. In general the Phoenicians and the Greeks as well as modern people founded their colonies in unoccupied localities. Here they raised up new towns which were located in places favorable to maritime and commercial relations. The Romans, on the contrary, avoided establishing colonies in

---

[1] Momm., *loc. cit.*
[2] Brutus (App. B. C., II, 140) calls the colonists, Φύλακας τῶν πεπολεμηκότων.
[3] Ihne, I, 236.
[4] Cicero, Ad Att., I, 19: "Sentinam urbis exhaurire, et Italiae solitudinem frequentori posse arbitrabor."

new places. When they had taken possession of a city, they expelled from it a part of the inhabitants, whether to transfer them to Rome as at first, or a little later, when it became necessary to discourage the increase of Roman population, to more distant places. The population thus expelled was replaced with Roman and Latin citizens.[1] Thus a permanent garrison was located which assured the submission of the neighboring countries and arrested in its incipiency every attempt at revolt. In every respect these colonies remained under surveillance and in a dependence the most complete and absolute upon the mother city, Rome. Colonies never became the means of providing for the impoverished and degraded until the time of Gaius Gracchus. When new territory was conquered, there went the citizen soldier. Thus these colonies mark the growth of Roman dominion as the circumscribed rings mark the annual growth of a tree. These colonies were of two kinds, Latin and Roman.

1. Latin colonies were those[2] which were composed of Latini and Hernici, or Romans enjoying the same rights as these, *i. e.* possessed of the Latin right rather than the Roman franchise. They were established inland as road fortresses and being located in the vicinity of mountain passes or main thoroughfares acted as a guard to Rome, and held the enemy in check.

2. Roman, or Burgess, colonies[3] were those composed wholly of Roman citizens who kept their political rights and consequent close union with their native city. In some cases Latini were given the full franchise and permitted to join these colonies. In position as well as rights, these colonies were distinguished from the Latin, being with few exceptions situated upon the coast and thus acting as guards against foreign invasion.

---

[1] Momm., I, 145.
[2] Marquardt u. Momm., IV, 35–51; Momm., *History of Rome*, I, 108, 539; Madvigi Opuscula Academica, I, 208–305.
[3] Marquardt u. Momm., IV, 35–51; Ihne, vols. I–V; Momm., vols. I–V; Madvigi Opus., *loc. cit.*

Table of Latin Colonies in Italy.

	COLONIES.	LOCATION.	B. C.	AUTHORITIES.
1	Signia.	Latium.	?	Livy, 1, 56; Dionys., 4, 63.
2	Cerceii.	"	?	Id.
3	Suessa Pometia.	"	?	Livy, 2, 16.
4	Cora.	"	?	Livy, 2, 16.
5	Velitrae.	"	494	Livy, 2, 30, 31; Dionys., 6, 42, 43.
6	Norba.	"	492	Livy, 2, 34; Dionys., 7, 13.
7	Antium.	"	467	" 3, 1; " 9, 59.
8	Ardea.	"	442	" 4, 11; Diodor., 12, 34.
9	Satricum.	"	385	" 6, 14.
10	Sutrum.	Etruria.	383	Vell., 1, 14.
11	Nepete.	"	383	Livy, 6, 21; Vell.
12	Setia.	Latium.	382	Vell., 1, 14; Livy, 6, 30.
13	Cales.	Campania.	334	" 1, 14; " 8, 16.
14	Fregellae.	Latium.	328	Livy, 8, 22.
15	Luceria.	Apulia.	314	" Epit., 60.
16	Suessa.		313	" 9, 28.
17	Pontiae.	Isle of Latium.	313	" 9, 28.
18	Saticula.	Samnium.	313	" 9, 22; Vell., 1, 14; Festus, p. 340.
19	Interamna Liriuas.	Latium.	312	Livy, 9, 28; Vell., 1, 14; Diodor., 19, 105.
20	Sora.	"	303	Livy, 10, 1; Vell., 1, 14.
21	Alba.	"	303	" 10, 1; " 1, 14.
22	Narnia.	Umbria.	299	" 10, 10.
23	Carseola.	Latium.	298	" 10, 13.
24	Venusia.	Apulia.	291	Vell., 1, 14; Dionys. Ex., 2335.
25	Hatria.	Picenum.	289	Livy, Epit., 11.
26	Cosa.	Campania.	273	" " 14; Vell., 1, 14.
27	Paestum.	Lucania.	273	Id. Id.
28	Ariminum.		268	Vell., 1, 14; L. Epit., 15; Eutrop., 2, 16.
29	Beneventum.	Samnium.	268	Vell., 1, 14; L. Epit., 15; Eutrop., 2, 16.
30	Firmum.	Picenum.	264	Vell., 1, 14.
31	Aesernia.	Samnium.	263	" 1, 14; L. Epit., 16.
32	Brundisium.	Calabria.	244	" 1, 14; " 19.
33	Spoletium.	Umbria.	241	" 1, 14; " 20.
34	Cremona.	Gallia Cis.	218	Tacitus, Hist., 3, 35.
35	Placentia.	" "	218	L. Epit., 20; Polyb., 3, 40; V. 1, 14, 8.
36	Copia.	Lucania.	193	Livy, 34, 53.
37	Valentia.	Bruttii.	192	" 34, 40; 35, 40.
38	Bononia.	Gallia Cis.	189	" 37, 57; Vell., 1, 15.
39	Aquileia.	Gallia Trans.	181	" 40, 34; " "

Table of Civic Colonies in Italy.

	COLONIES.	LOCATION.	B. C.	AUTHORITY.
1	Ostea.	Latium.	418	Livy, 1, 33; Dionys., 3, 44; Polyb., 6, 29; Cic. de R. R., 2, 18, 33.
2	Labici.	"	418	Livy, 4, 47, 7.
3	Antium.	"	338	" 8, 14.
4	Auxur.	"	329	" 8, 21; 27, 38; Vell., 1, 14.
5	Minturnae.	Campania.	296	Livy, 10, 21.
6	Sinuessa.	"	296	" 10, 21; 27, 38.
7	Sena Gallica.	Umbria.	283	" Epit., 11; Vell., 1, 14, 8.
8	Castrum Novum.	Picenum.	283	Livy, Epit., 11; Vell., 1, 14, 8.
9	Aesium.	Umbria.	247	Vell., 1, 14, 8.
10	Alsium.	Etruria.	247	" 1, 14, 8; L. Epit., 19; L., 36, 3.
11	Fregena.	"	245	Livy, 36, 3.
12	Pyrgi.	"	191	" "
13	Puteoli.	Campania.	194	" 34, 45.
14	Volturnum.	"	194	Id.
15	Liturnum.	"	194	Id.
16	Salernum.	"	194	Id.
17	Buxentum.	Lucania.	194	Livy, 34, 45.
18	Sipontum.	Apulia.	194	Id.
19	Tempsa.	Bruttii.	194	Id.
20	Croton.	"	194	Id.
21	Potentia.	Picenum.	184	Livy, 39, 44.
22	Pisaurum.	Umbria.	184	" " "
23	Parma.	Gallia Cis.	183	" " 55.
24	Mutina.	Gallia Cis.	183	Livy, 39, 55.
25	Saturnia.	Etruria.	183	" " "
26	Graviscae.	"	181	" 40, 39.
27	Luna.	"	180	" 41, 13.
28	Auximum.	Picenum.	157	Vell., 1, 15, 3.
29	Fabrateria.	Latium.	124	" 1, 15, 4.
30	Minervia.	Bruttii.	122	" 1, 15, 4; Appian B. C., 2, 23.
31	Neptunia.	Iapygia.	122	Id.
32	Dertona.	Liguria.	100	Vell., 1, 15, 5.
33	Eporedia.	Gallia Trans.	100	" " "
34	Narbo Martius.	" Narbo.	118	Mommsen.

CHAPTER II.

Sec. 5.—Lex Cassia.

Every year added to the difference between the patrician and plebeian, the rich and the poor; a difference which had now grown so great as to threaten seriously the very existence of the state. The most sagacious of all the plans which had been proposed to stop this evil, was that set forth by Spurius Cassius, a noble patrician now acting as consul for the third [1] time. In the year 268, he submitted to the burgesses [2] a proposal to have the public land surveyed, that portion belonging to the populus set aside and the remainder divided among the plebeians or leased for the benefit [3] of the public treasury.

He thus attempted to wrest from the senate the control of the public land and, with the aid of the Latini and the plebeians, to put an end to the system of occupation.[4] The lands which he proposed to divide were solely those which the state had acquired through conquest since the general assignment by king Servius, and which it still retained.[5] This was the first measure by which it was proposed to disturb the possessors in their peaceful occupation of the state lands, and, according to Livy, such a measure had never been proposed from then to the time in which he was writing, under Augustus, without exciting the greatest disturbance.[6] Cassius might well

---

[1] Dionysius, VIII, 68; "Οἱ δέ παρὰ τούτων τὴν ὑπατείαν παραλαβόντης πόπλιος Οὐεργίνιος καὶ Σπόριος Κάσσιος, τὸ τρίτον τότε ἀποδειχθεὶς ὕποτος, κ. τ. λ."
[2] Dionysius, VIII, 69; Livy, II, 41, seq.   [3] Dionysius, VIII, 81.
[4] Dionysius, VIII, 69; Mommsen, I, 363.   [5] Niebuhr, II, 166.
[6] Livy, II, 41; "Tum primum lex agraria promulgata est nunquam deinde usque ad hanc memoriam sine maximus motibus rerum agitata."

suppose that his personal distinction and the equity and wisdom of the measure would carry it through, even amidst the storm of opposition to which it was subjected. Like many other reformers equally well meaning, he was mistaken.

The citizens who occupied this land had grown rich by reason of its possessions. Some of them received it as an inheritance, and doubtless looked upon it as their property as much as the *Ager Romanus*. These to a man opposed the bill. The patricians arose en masse. The rich plebeians, the aristocracy of wealth, took part with them. Even the commons were dissatisfied because Spurius Cassius proposed in accordance with federal rights and equity to bestow a portion of the land upon the Latini and Hernici, their confederates and allies.[1] The bill proposed by Cassius, together with such provisions as were necessary, became a law, according to Niebuhr,[2] because the tribunes had no power to bring forward a law of any kind before the plebeian tribes obtained a voice in the legislature by the enactment of the Publilian law in 472 B. C.: so that when they afterwards made use of the agrarian law to excite the public passions it must have been one previously enacted but dishonestly set aside and, in Dionysius' account, this is the form which the commotion occasioned by it takes.[3] Though this is doubtless true, yet the law, by reason of the combined opposition, became a dead letter and the people who would have been most benefited by its enforcement joined with Cassius' enemies at the expiration of his term of office to condemn him to death. In this way does ignorance commonly reward its benefactors. This agitation aroused by Cassius, stirred the Roman Commonwealth, now more than twenty years old, to its very foundations, but it had no immediate effect upon the *ager publicus*. The rich patrician together with the few plebeians who had wealth

---

[1] Livy, II, 41; Dionysius, VIII, 69.   [2] Niebuhr, II.
[3] Dionysius, VIII, 81: "ἐκκλησίαί τε συνεχεῖς ὑπὸ τῶν τότε δημάρχων ἐγίνοντο καὶ ἀπαιτήσεις τῆς ὑποσχέσεως." See also VIII, 87, line 25 et seq.

enough to farm this land, still held undisputed possession. The poor plebeian still continued to shed his blood on the battle field to add to Roman territory, but no foot of it did he obtain. Wealth centralized. Pauperism increased.

SEC. 6.—AGRARIAN MOVEMENTS BETWEEN 486 AND 367.

Modern historians who have written upon the Roman Republic have, so far as I know, passed immediately from the consideration of the *Lex Cassia* to the law of Licinius Stolo. Meanwhile more than a century had passed away. Cassius died in 485, Licinius Stolo proposed his law in 376. During this century which had beheld the organization of the republic and the growth, by tardy processes, of the great plebeian body many agrarian laws were proposed and numerous divisions of the public land took place. Both Dionysius and Livy mention them. The poor success of the proposition of Cassius and the evil consequences to himself in no way checked the zeal of the tribunes. Propositions of agrarian laws followed one another with wonderful rapidity. Livy enumerates these propositions, but almost wholly without detail and without comments upon their tendencies or points of difference from one another or from the law of Cassius. As this law failed of its object by being disregarded, we may safely conclude that the most of these propositions were but a reproduction of the law of Cassius.

In 484, and again in 483, the tribune proposed agrarian laws but what their nature was, Livy, who records them, does not tell us. From some vague assertions which he makes we may conclude that the point of the law was well known, and was but a repetition of that of Cassius.[1] The consul Caeso Fabius, in 484, and his brother Marcus in the

---

[1] "Solicitati, eo anno, sunt dulcedine agrariae legis animi plebis, . . . vana lex vanique legis auctores." Livy, II, 42.

following year, secured the opposition of the senate and succeeded in defeating their laws.

Livy (II, 42,) mentions very briefly a new proposition brought forward by Spurius Licinius in 482. Here we are able to complete his account by reference to Dionysius,[1] who says that, in 483, a tribune named Caius Maenius had proposed an agrarian law and declared that he would oppose every levy of troops until the senate should execute the law ordaining the creation of decemvirs to determine the boundaries of the domain land and, in fine, forbid the enrolment of citizens. The senate was able through the consuls, Marcus Fabius and Valerius, the ancient colleague of Cassius, to invent a means of avoiding this difficulty. The authority of the tribunes by the old Roman law,[2] did not reach without the walls of the city, while that of the consuls was everywhere equal and only bounded by the limits of the Roman world. They moved their curule chairs and other insignia of their authority without the city walls and proceeded with the enrolments. All who refused to enroll were treated as enemies[3] of the republic. Those who were proprietors had their property confiscated, their trees cut down, and their houses burned. Those who were merely farmers saw themselves bereft of their farm-implements, their oxen and all things necessary for the cultivation of the soil. The resistance of the tribunes was powerless against this systematic oppression on the part of the patricians; the agrarian[4] law failed and the enrolment progressed.

There is some difficulty in determining the facts of the law proposed by Spurius Licinius[5] of which Livy speaks. Dionysius calls this tribune, not Licinius but Σπύριος Σικίλιος. The Latin translation of Dionysius has the name Icilius and this has been the name adopted by Sigonius and other his-

---

[1] Dionysius, VIII, 606, 607.
[2] Livy, *loc. cit.*: Dionysius, *loc. cit.*
[3] Dionys., VIII, 554.
[4] Dionys., VIII, 555.
[5] Val. Max., Fg. of Bk. X : " Spurii, patre incerto geniti."

torians. Livy tells us that the Icilian family was at all times hostile to the patricians and mentions many tribunes by this name who were staunch defenders of the commons. In accepting this correction, therefore, it is not necessary to confound this Icilius with the one who proposed the partition of the Aventine among the plebeians. Icilius, according to both Livy and Dionysius,[1] made the same demand as the previous tribunes, *i. e.*, that the decemvirs should be nominated for the survey and distribution of the domain lands, according to previous enactment. He further declared that he would oppose every decree of the senate either for war or the administration of the interior until the adoption and execution of his measures. Again the senate avoided the difficulty and escaped, by a trick, the execution of the law. Appius Claudius, according to Dionysius,[2] advised the senate to search within the tribunate for a remedy against itself, and to bribe a number of the colleagues of Icilius to oppose his measure. This political perfidy was adopted by the senate with the desired effect. Icilius persisted in his proposition and declared he would rather see the Etruscans masters of Rome than to suffer for a longer time the usurpation of the domain lands on the part of the possessors.[3]

This somewhat circumstantial account has revealed to us that at this time it took a majority of the tribunes to veto an act of their colleague. At the time of the Gracchi the veto of a single tribune was sufficient to hinder the passage of a law, and Tiberius was for a long time thus checked by his colleague, Octavius. Then the tribunician college consisted of ten members, and it would be no very difficult thing to detach one of the number either by corruption or jealousy. But it is evident that, at the time we are considering, it took a majority of the tribunes to veto an act of a colleague;

---

[1] Livy, *loc. cit.*; Dionys., *loc. cit.*   [2] Dionys., IX, 558; Livy, II, 43.
[3] Dionys., IX, 559–560: "τοὺς κατέχοντος τὴν χώραν τὴν δημοσίαν." . . . "Καὶ Σικίλιος οὐδενὸς ἔτι κύριος ἦν."

moreover, the college consisted of five members. This latter fact is seen in the statement of Livy,[1] when he mentions the opposition which four of the tribunes offered to their colleague, Pontificius, in 480. In this same case he attributes to Appius Claudius the conduct which Dionysius attributed to him in the previous year. But he causes Appius to state, in his speech favoring the corruption of certain tribunes, "that the veto of one tribune would be sufficient to defeat all the others."[2] This is contrary to the statement of Dionysius[3] and would seem improbable, for, if the opposition of one tribune was sufficient, the patricians would not have deemed it necessary to purchase four. That would be contrary to political methods.

Of the two propositions of the tribunes, Icilius, in 482, and Pontificius, in 480, the results were the same. The opposition of their colleagues defeated them. But this persistent opposition rather than crushing seemed to stir up renewed attacks. We have seen the tribunes, Menius, Icilius, and Pontificius, successively fail. The next movement was led by a member of the aristocracy, Fabius Caeso,[4] consul for the third time in 477. He undertook to remove from the hands of the tribunes the terrible arm of agrarian agitation which they wielded constantly against the patricians, by causing the patricians themselves to distribute the domain lands equally among the plebeians, saying : "that those[5] persons ought to have the lands by whose blood and sweat they had been gained." His proposition was rejected with scorn by the patricians, and this attempt at reconciliation failed as all the attempts of the tribunes had. The war

---

[1] Livy, loc. cit.
[2] Livy, II, 44: "Et unum vel adversus omnes satis esse . . . quatuorque tribunorum adversus unum."
[3] Dionys., IX, 562. [4] Livy, loc. cit.; Dionys., loc. cit.
[5] Livy, II, 48: "Captivum agrum plebi, quam maxime aequaliter darent. Verum esse habere eos quorum sanguine ac sudore partus sit. Aspernati Patres sunt."

with Vaii which, according to Livy, now took place hindered for a while any agrarian movements; but, in 474, the tribunes Gaius Considius and Titus Genucius made a fruitless attempt at distribution, and, in 472, Dionysius speaks of a bill brought forward by Cn. Genucius which is probably the same bill.

In 468, the two consuls, Valerius and Aemilius, faithfully supported the tribunes in their demand[1] for an agrarian law. The latter seems to have supported the tribunes because he was angry that the senate had refused to his father the honor of a triumph; Valerius, because he wished to conciliate the people for having taken part in the condemnation of Cassius.

Dionysius, according to his custom, takes advantage of the occasion to write several long speeches here, and one of them is valuable to us. He causes the father of Aemilius to set forth in a formal speech the true character of the agrarian laws and the right of the state to again assume the lands which had been taken possession of. He further says: "that it is a wise policy[2] to proceed to the division of the lands in order to diminish the constantly increasing number of the poor, to insure a far greater number of citizens for the defense of the country, to encourage marriages, and, in consequence, to increase the number of children and defenders of the republic." We see in this speech the real purpose, the germ, of all the ideas which Licinius Stolo, the Gracchi, and even Cæsar, strove to carry out. But the Roman aristocracy was too blind to comprehend these words of wisdom. All these propositions were either defeated or eluded.

Lex Icilia. In the year 454,[3] Lucius Icilius, one of the tribunes for that year, brought forward a bill that the Aventine hill should be conveyed to the plebeians as their personal and

---

[1] Livy, II, 61, 63, 64.
[2] Dionys., IX, 606, 607; Livy, III, 1.
The authorities are somewhat conflicting at this point, and I have followed the account of Dionysius.
[3] Schwegler, *Römische Geschichte*, II, 484; Dionys., X, 31, p. 657, 43.

especial property.¹ This hill had been the earliest home of the plebeians, yet they had been surrounded by the lots and fields of the patricians. That part of the hill which was still in their possession was now demanded for the plebeians. It was a small thing for the higher order to yield this much, as the Aventine stood beyond the Pomoerium,² the hallowed boundary of the city, and, at best, could not have had an area of more than one-fourth of a square mile, and this chiefly woodland. The consuls, accordingly, made no hesitation about presenting the bill to the senate before whom Icilius was admitted to speak in its behalf. The bill was accepted by the senate and afterwards confirmed by the Centuries.³ The law provided,—" that all the ground which has been justly acquired by any persons shall continue in the possession of the owners, but that such part of it as may have been usurped by force or fraud by any persons and built upon, shall be given to the people; those persons being repaid the expenses of such buildings by the estimation of umpires to be appointed for that purpose, and that all the rest of the ground belonging to the public, be divided among the people, they paying no consideration for the same."⁴ When this was done the plebeians took possession of the hill with solemn ceremonies. This hill did not furnish homes for all the plebeians, as some have held; nor, indeed, did they wish to leave their present settlements in town or country to remove to the Aventine. Plebeians were already established in almost all parts of the city and held, as vassals of the patricians, considerable portions of Roman territory. This little hill could never have furnished⁵ homes of any sort to the whole plebeian popula-

---

¹ Dionys., X, 31, l. 13; Ihne, *Hist. of Rome*, 1, 191, note; Lange, *Röm. Alter.*, I, 619. Also see art. in Smith's *Dict. of Antiquities*.
² *I. e.* outside of the '*quadrata*' but ἐμπεριεχόμενος τῇ πόλεις, Dionys., X, 31, l. 18: "pontificale pomoerium, qui auspicato olim quidem omnem urbem ambiebat praeter Aventinum." Paul. ex Fest., p. 248, Müll.
³ Dionys., X, 32. ⁴ Dionys., X, 32. ⁵ Momm., I, 355.

tion. What it did do was to furnish to the plebeians a trysting place in time of strife with their patrician neighbors, where they could meet, apart and secure from interruption, to devise means for resisting the encroachments of the patricians and to further establish their rights as Roman citizens. Thus a step toward their complete emancipation was taken. For a moment the people were soothed and satisfied by their success, but soon they began to clamor for more complete, more radical, more general laws. An attempt seems to have been made in 453 to extend the application of the *lex Icilia* to the *ager publicus*,[1] in general, but nothing came of it. In 440, the tribune, Petilius, proposed an agrarian law. What its conditions were Livy has not informed us, but has contented himself with saying that "Petilius made a useless attempt to bring before the senate a law for the division of the domain lands."[2] The consuls strenuously opposed him and his effort came to naught.

In our review of the agrarian agitation we must mention the forceless and insignificant attempt made by the son of Spurius Melius, in 434. Again, in 422, we find that other attempts were made which availed nothing. Yet the tribunes who attempted thus to gain the good will of the people set forth clearly the object which they had in view in bringing forward an agrarian bill. Says Livy; "They held out the hope to the people of a division of the public land, the establishment of colonies, the levying of a *vectigal* upon the possessors, which *vectigal* was to be used[3] in paying the soldiers."

In the year 419, and again in 418, unavailing attempts were made for the division of lands among the plebeians.

---

[1] Dionys., X, 34.

[2] Livy, IV, 12: Neque ut de agris dividendis plebi referrent consules ad senatum pervincere potuit. . . . Ludibrioque erant minae tribuni.

[3] "Agri publici dividendi, coloniarumque deducendarum ostentatae spes, et vectigali possessoribus imposito, in stipendium militum erogandi aeris." Livy, IV, 36.

Spurius Maecilius and Spurius Metilius, the tribunes[1] for the year 412, proposed to give to the people, in equal lots, the conquered lands. The patricians ridiculed this law, stating that Rome itself was founded upon conquered soil and did not possess a single acre of land that had not been taken by force of arms, and that the people held nothing save that which had been assigned by the republic. The object, then, of the tribunes was to distribute the fortunes of the entire state. Such vapid foolishness as this failed not of the effect which the patricians aimed at. Appius Claudius counselled the adoption of the excellent means invented by his grandfather. Six tribunes were bought over by the caresses, flatteries, and money of the patricians and opposed their vetoes to their colleagues who were thus compelled to retire.[2]

In the following year, 411, Lucius Sextius, in no way discouraged by the ill success of his predecessors, proposed the establishment of a colony at Bolae, a town in the country of the Volscians, which had been recently conquered. The patricians[3] opposed this by the same method which they had adopted in the preceding case, the veto by tribunes. Livy criticises the impolitic opposition of the patricians in these words : " This was a most seasonable time, after the punishment of the mutiny, that the division of the territory of Bolae should be presented as a soother to their minds ; by which proceeding they would have diminished their eagerness for an agrarian law, which tended to expel the patricians from the public land unjustly possessed by them. Then this very indignity exasperated their minds, that the nobility persisted not only in retaining the public lands, which they got possession of by force, but would not even grant to the commons the unoccupied land lately taken from the enemy, and which would, like the rest,[4] soon become the prey of the few."

---

[1] Livy, loc. cit.
[2] Livy, IV, 48.
[3] Livy, IV, 49.
[4] Livy, IV, 51.

In 409, Icilius, without doubt a member of that plebeian
family which had furnished so many stout defenders of the
liberties of the people, was elected tribune of the people
and brought forward an agrarian bill, but a plague broke
out and hindered any further action. In 407, the tribune,
Menius, introduced an agrarian bill and declared that he
would oppose the levies until the persons who unjustly
held the public domains consented to a division. A war
broke out and agrarian legislation was drowned amid the din
of arms. Some years now elapsed without the mention of
any agrarian laws. The siege of Veii commenced in 406
and lasted for six years, during which time military law
was established, giving occupation and some sort of satisfac-
tion to the plebeians. In 397, an agrarian movement was set
on foot, but the plebeians were partially satisfied by being
allowed to elect one of their number as *tribunus consularis* for
the following year, thus obtaining a little honor but no land.
After the conquest of Veii, there was a movement on the
part of the plebeians to remove from Rome and settle upon
the confiscated territory of the Veians; this was only staid
by concessions on the part of the patricians. A decree of the
senate was passed,—"that seven jugera, a man, of Vientian
territory, should be distributed to the commons and not only
to the fathers of families, but also that all persons in their
house in the state of freedom should be considered, and that
they might be willing to rear up children[1] with that prospect."
In 384, six years after the conquest of Rome by the Gauls,
the tribunes of the year proposed a law for the division of the
Pomptine territory (*Pomptinus Ager*) among the plebeians.
The time was not a favorable one for the agitation of the
people, as they were busy with the reconstruction of their
houses laid waste by the Gauls, and the movement came to
nothing. . The tribune, Lucius Licinius, in 383, revived this
movement but it was not successfully carried till the year 379,

---

[1] Livy, VI, 5.

when the senate, well disposed towards the commons by reason of the conquest of the Volscians, decreed the nomination of five commissioners to divide the Pomptine territory[1] among the plebs. This was a new victory for the people and must have inspired them with the hope of one day obtaining in full their rights in the public domains.

We have now passed in review the agrarian laws proposed and, in some cases, enacted between the years 485 and 376, *i. e.* between the *lex Cassia* and the *lex Licinia*, which the greater part of the historians have neglected. We have now come to the propositions of that illustrious plebeian whose laws, whose character, and whose object have been so diversely appreciated by all those persons who have studied in any way the constitutional history of Rome. We wish to enter into a detailed examination of the *lex Licinia*, but before so doing have deemed it expedient to thus pass in review the agrarian agitations. The result of this work has, we trust, been a better understanding of the real tendency, the true purpose, of the law which is now to absorb our attention. It was no innovation, as some writers of the day assert, but in reality confined itself to the well beaten track of its predecessors, striving only to make their attainments more general, more substantial and more complete.

Extension of Territory by Conquest up to the Year 367 B. C.

1. Corcoli, captured in 442.
2. Bolae, captured in 414.
3. Labicum, captured in 418.
4. Fidenae, captured in 426 and all the territory confiscated.
5. Veii, captured in 396. This was the chief town of the Etruscans, equal to Rome in size, with a large tributary country; territory confiscated.

Approximate amount of land added to the Roman domain, 150 square miles.

---

[1] Quinque viros Pomptino agro dividendo. Livy, VI, 21.

Colonies Founded between 454 and 367.

CIVIC COLONIES.

COLONIES.	PLACE.	DATE.	NO. OF COLONISTS.	NO. OF JUG. TO EACH.	TOTAL NO. OF JUG.	ACRES.
Labici.	Latium.	418	1500	2	3000	1875

LATIN COLONIES.

Ardea.	Latium.	442	300	2	600	375
Satricum.	"	385	300	2	600	375
Sutrium.	Etruria.	383	300	2	600	375
Nepete.	"	383	300	4	1200	750
Setia.	Latium.	382	300	4	1200	750
				Total......	7200	4500

SEC. 7.—LEX LICINIA.

Party lines were, at the time of the enactment of the Licinian Law, strongly marked in Rome. One of the tribunes chosen after the return of the plebeians from Mons Sacer was a Licinius. The first military tribune with consular power elected from the plebeians was another Licinius Calvus. The third great man of this distinguished family was Caius Licinius Calvus Stolo, who, in the prime of life and popularity, was chosen among the tribunes of the plebs for the seventh year following the death of Manlius the Patrician. Another plebeian, Lucius Sextius by name, was chosen tribune at the same time. If not already, he soon became the tried friend of Licinius. Sextius was the younger but not the less earnest of the two. Both belonged to that portion of the plebeians supposed to have been latterly connected with the liberal patricians. The more influential and by far the more reputable members of the lower estate were numbered in this party. Opposed to it were two other parties of plebeians. One consisted of the few who, rising to wealth or rank, cast off the bonds uniting them to the lower estate. They preferred to be upstarts among patricians rather than leaders among plebeians. As a matter of course, they became the parasites of the illiberal patricians. To the same body was attached another plebeian party. This was formed of the inferior classes belonging to the lower estate. These inferior plebeians were generally disregarded by the higher classes of their own estate as well as by the patricians of both the liberal and illiberal parties. They were the later comers, or the poor and degraded among all. As such they had no other resource but to depend on the largesses or the commissions of the most lordly of the patricians. This division of the plebeians is a point to be distinctly marked. While there were but two parties, that is the liberal and the illiberal among the patricians, there were no less than three among the plebeians. Only one of the three could be called a plebeian party. That was the party containing the nerve and sinew of the order, which united only with the liberal patricians, and with them only on comparatively independent terms. The other two parties were nothing but servile retainers of the illiberal patricians.

It was to the real plebeian party that Licinius belonged, as also did his colleague Sextius,[1] by birth. A tradition of no value represented the patrician and the plebeian as being combined to support the same cause in consequence of a whim of the wife and daughter through whom they were connected. Some revolutions, it is true, are the effect of an instant's passion or an hour's weakness. Nor can they then make use of subsequent achievements to conceal the caprices or the excitements in which they originated. But a change, attempted by Licinius with the help of his father-in-law, his colleague, and a few friends reached back one hundred years and more (B. C. 486) to the law of the martyred Cassius, and

---

[1] Livy, VI, 34.

forward to the end of the Commonwealth. It opened new honors as well as fresh resources to the plebeians.

Probably the tribune was raised to his office because he had shown the determination to use its powers for the good of his order and of his country. Licinius and Sextius together brought forward the three bills bearing the name of Licinius as their author. One, says the historian, ran concerning debts. It provided that, the interest already[1] paid being deducted from the principal, the remainder should be discharged in equal installments within three years. The statutes against excessive rates of interest, as well as those against arbitrary measures of exacting the principal of a debt, had utterly failed. It was plain, therefore, to any one who thought upon the matter,— in which effort of thought the power of all reformers begins, —that the step to prevent the sacrifice of the debtor to the creditor was still to be taken. Many of the creditors themselves would have acknowledged that this was desirable. The next bill of the three related to the public lands. It prohibited any one from occupying more than five hundred jugera, about 300 acres; at the same time it reclaimed all above that limit from the present occupiers, with the object of making suitable apportionments among the people[2] at large. Two further clauses followed, one ordering that a certain number of freemen should be employed on every estate; another forbidding any single citizen[3] to send out more than a hundred of the larger, or five hundred of the smaller cattle to graze upon the public pastures. These latter details are important, not so much in relation to the bill itself

---

[1] Livy, VI, 35: "unam de aere alieno, ut deduco eo de capite, quod usuris pernumeratum esset, id, quod superesset, triennio aequis portionibus persolveretur."

[2] Livy, VI, 35; Niebuhr, III, p. 16; Varro, De R. R., 1: "Nam Stolonis illa lex, quae vetat plus D jugera habere civem Romanorum." Livy, VI, 35: "alteram de modo agrorum, ne quis plus quingenta jugera agri possideret." Marquardt u. Momm., Röm. Alterthümer, IV, S. 102.

[3] Appian, De Bello Civile, I, 8.

as to the simultaneous increase of wealth and slavery which they plainly signify. As the first bill undertook to prohibit the bondage springing from too much poverty, so the second aimed at preventing the oppression springing from too great opulence. A third bill declared the office of military tribune with consular power to be at an end. In its place the consulate was restored with full[1] provision that one of the two consuls should be taken from the plebeians. The argument produced in favor of this bill appears to have been the urgent want of the plebeians to possess a greater share in the government than was vested in their tribunes, aediles, and quaestors. Otherwise, said Licinius and his colleague, there will be no security that our debts will be settled or that our lands will be obtained.[2] It would be difficult to frame three bills, even in our time, reaching to a further, or fulfilling a larger reform. "Everything was pointed against the power of the patricians[3] in order to provide for the comfort of the plebeians." This to a certain degree was true. It was chiefly from the patrician that the bill concerning debts detracted the usurious gains which had been counted upon. It was chiefly from him that the lands indicated in the second bill were to be withdrawn. It was altogether from him that the honors of the consulship were to be derogated. On the other hand the plebeians, save the few proprietors and creditors among them, gained by every measure that had been proposed. The poor man saw himself snatched from bondage and endowed with an estate. He who was above the reach of debt saw himself in the highest office of the state. Plebeians with reason exulted. Licinius evidently designed reuniting the divided members of the plebeian body. Not one of them, whether rich or poor, but seems called back by these bills to stand with his own order from that time on.

---

[1] Livy, VI, 35; See Momm., I, 382; Duruy, *Hist. des Romains*, II, 78.
[2] Livy, VI, 37.
[3] Livy, VI, 35: "creatique tribuni Caius Licinius et Lucius Sextius promulgavere leges adversus opes patriciorum et pro commodis plebis."

If this supposition was true, then Licinius was the greatest leader whom the plebeians ever had up to the time of Cæsar. But[1] from the first he was disappointed. The plebeians who most wanted relief cared so little for having the consulship opened to the richer men of their estate that they would readily have dropped the bill concerning it, lest a demand should endanger their own desires. In the same temper the more eminent men of the order, themselves among the creditors of the poor and the tenants of the domain, would have quashed the proceedings of the tribunes respecting the discharge of debt and the distribution of land, so that they carried the third bill only, which would make them consuls without disturbing their possessions. While the plebeians continued severed from one another, the patricians drew together in resistance to the bills. Licinius stood forth demanding, at once, all that it had cost his predecessors their utmost energy to demand, singly and at long intervals, from the patricians. Nothing was to be done but to unite in overwhelming him and his supporters. "Great things were those that he claimed and not to be secured without the greatest contention."[2] The very comprehensiveness of his measures proved the safeguard of Licinius. Had he preferred but one of these demands, he would have been unhesitatingly opposed by the great majority of the patricians. On the other hand he would have had comparatively doubtful support from the plebs. If the interests of the poorer plebeians alone had been consulted, they would not have been much more active or able in backing their tribunes, while the richer men would have gone over in a body to the other side with the public tenants and the private creditors among the patricians. Or, supposing the case reversed and the bill relating to the consulship brought forward alone, the debtors and the homeless citizens would have given the bill too little help with

---

[1] Ihne, I, 314.
[2] Livy, VI, 35: "Cuncta ingentia, et quae sine certamine obtineri non possent."

hands or hearts to secure its passage as a law. The great encouragement therefore to Licinius and Sextius must have been their conviction that they had devised their reform on a sufficiently expanded scale. As soon as the bills were brought forward every one of their eight colleagues vetoed their reading. Nothing could be done by the two tribunes except to be resolute and watch for an opportunity for retaliation. At the election of the military tribunes during that year, Licinius and Sextius interposed[1] their vetoes and prevented a vote being taken. No magistrates could remain in office after their terms expired, whether there were any successors elected or not to come after them. The commonwealth remained without any military tribunes or consuls at its head, although the vacant places were finally filled by one *interrex* after another, appointed by the senate to keep up the name of government and to hold the elections the moment the tribunes withdrew their vetoes, or left their office. At the close of the year Licinius and Sextius were both re-elected but with colleagues on the side of their antagonists. Some time afterwards it became necessary to let the other elections proceed. War was threatening,[2] and in order to go to the assistance of their allies Licinius and Sextius withdrew their vetoes and ceased their opposition for a time. Six military tribunes were chosen, three from the liberal and three from the illiberal patricians. The liberals doubtless received all the votes of the plebeians as they had no candidates. They had in all probability abstained from running for an office, bills for the abolition of which were held in abeyance. They showed increasing inclination to sustain Licinius and his colleague, both by re-electing them year after year and by at length choosing three other tribunes with them in favor of the bills. The prospects of the measure were further brightened by the election of Fabius Ambustus, the father-in-law of Licinius and his zealous sup-

---

[1] Livy, VI, 35.   [2] Livy, VI, 36.

porter, to the military[1] tribunate. This seems to have been the seventh year following the proposal of the bills. This can not be definitely determined, however. During this long period of struggle, Licinius had learned something. It was constantly repeated[2] in his hearing that not a plebeian in the whole estate was fit to take the part in the auspices and the religious ceremonies incumbent upon the consuls. The same objections had overborne the exertions of Caius Canuleius three-quarters of a century before. Licinius saw that the only way to defeat this argument was by opening to the plebeians the honorable office of *duumvirs*, whose duty and privilege it was[3] to consult the Sibyline books for the instruction of the people in every season of doubt and peril. They were, moreover, the presiding officers of the festival of Apollo, to whose inspirations the holy books of the Sibyl were ascribed, and were looked up to with honor and respect. This he did by setting forth an additional bill, proposing the election of *decemvirs*.[4] The passage of this bill would forever put to rest one question at least. Could he be a decemvir, he could also be a consul. This bill was joined to the other three which were biding their time. The strife went on. The opposing tribunes interposed their vetoes. Finally it seems that all the offices of tribune were filled with partisans of Licinius, and the bills were likely to pass when Camillus, the dictator, swelling with wrath against bills, tribes and tribunes,[5] came forward into the forum. He commanded the tribunes to see to it that the tribes cast no more votes. But on the contrary they ordered the people to continue as they had begun. Camillus ordered his lictors to break up the assembly and proclaim that if a man lingered in the forum, the dictator would call out every man fit for service and march

---

[1] Livy, VI, 36. Fabius quoque tribunis militum, Stolonis socer, quarum legum auctor fuerat, earum sua.
[2] Livy, *loc. cit.*   [3] Appian, *De Bell. Civ.*, I, 9.
[4] Momm., I, 240: "decemviri sacris faciundis." Lange, *loc. cit.*
[5] Livy, VI, 38; Momm., *loc. cit.*

from Rome. The tribunes ordered resistance and declared that if the dictator did not instantly recall his lictors and retract his proclamation, they, the tribunes, would, according to their right, subject him to a fine five times larger than the highest rate of the census, as soon as his dictatorship expired. This was no idle threat, and Camillus retreated so fairly beaten as to abdicate immediately under the pretense of faulty auspices.[1] The plebeians adjourned satisfied with their day's victory. But before they could be again convened some influence was brought to bear upon them so that when the four bills were presented only the two concerning land and debts were accepted. This was nothing less than a fine piece of engineering on the part of the patricians to defeat the whole movement and could have resulted in nothing less. Licinius was disappointed but not confounded. With a sneer at the selfishness as well as the blindness of those who had voted only for what they themselves most wanted he bade them take heed that they could not eat if they would not drink.[2] He refused to separate the bills. The consent to their division would have been equivalent to consenting to the division of the plebeians. His resolution carried the day. The liberal patricians as well as the plebeians rallied to his support. A moderate patrician, a relation of Licinius, was appointed dictator, and a member of the same house was chosen master of the horse. These events prove that the liberal patricians were in the majority. Licinius and Sextius were re-elected for the tenth time, A. C. 366, thus proving that the plebeians had decided to eat and drink.[3]

The fourth bill, concerning the decemvirs was almost instantly laid before the tribes and carried through them. It was accepted by the higher assemblies and thus became a law. It is not evident why this bill was separated from the

---

[1] Livy, VI, 38; Momm., *loc. cit.*
[2] Dion Cassius, Fragment, XXXIII, with Reimer's note.
[3] Livy, VI, 42.

others, especially when Licinius had declared that they should not be separated. Possibly it was to smooth the way for the other three more weighty ones, especially the bill concerning the consulship.[1] There seems to have been an interruption here caused by an invasion of the Gauls.[2] As soon as this was over the struggle began again. The tribes assembled. " Will you have our bills?" asked Licinius and Sextius for the last time. " We will," was the reply. It was amid more violent conflicts, however, than had yet arisen that the bills became laws[3] at last.

It takes all the subsequent history of Rome to measure the consequences of the Revolution achieved by Licinius and Sextius; but the immediate working of their laws could have been nothing but a disappointment to their originators and upholders. We can tell little or nothing about the regard paid to the *decemvirs*. The priestly robes must have seemed an unprecedented honor to the plebeian. For some ten years the law regarding the consulship was observed, after which time it was occasionally[4] violated, but can still be called a success. The laws[5] of relief, as may be supposed of all such sumptuary enactments, were violated from the first. No general recovery of the public land from those occupying more than five hundred[6] jugera ever took place. Consequently there was no general division of land among the lackland class. Conflicting claims and jealousy on the part of the poor must have done much to embarrass and prevent

---

[1] Livy, VI, 42: et comitia consulum adversa nobilitate habita, quibus Lucius Sextius de plebe primus consul factus.
[2] Livy, *loc. cit.*
[3] Livy, VI, 42; Ovid, Faustus, I, 641, seq.:

" Furius antiquam populi superator Hetrusci
Voverat et voti solverat ante fidem
Causa quod a patribus sumtis secesserat annis
Vulgus; et ipsa suas Roma timebat opes."

[4] Momm., I, 389.   [5] Momm., I, 384.
[6] Arnold, *Roman History*, II, 35; Ihne, *Essay on the Roman Constitution*, p. 72. Ihne, *Roman Hist.*, I, 332-334. Long, I, ch. XI. Lange, *loc. cit.*

the execution of the law. No system of land survey to distinguish between *ager publicus* and *ager privatus* existed. Licinius Stolo himself was afterwards convicted of violating his own law.[1] The law respecting debts met with much the same obstacles. The causes of embarrassment and poverty being much the same and undisturbed, soon reproduced the effects which no reduction of interest or installment of principal could effectually remove. It is not our intention, however, to express any doubt that the enactments of Licinius, such as they were, might and did benefit the small farmer and the day laborer.[2] Many were benefited. In the period immediately following the passing of the law, the authorities watched with some interest and strictness over the observance of its rules and frequently condemned the possessors of large herds and occupiers of public domain to heavy fines.[3] But in the main the rich still grew richer and the poor and mean, poorer and more contemptible. Such was ever the liberty of the Roman. For the mean and the poor there was no means of retrieving their poverty and degradation.

These laws, then, had little or no effect upon the domain question or the re-distribution of land. They did not fulfil the evident expectation of their author in uniting the plebeians into one political body. This was impossible. What they did do was to break up and practically abolish the patriciate.[4] Henceforth were the Roman people divided into rich and poor only.

---

[1] Livy, VII, 16: "Eodem anno Caius Licinius Stolo a Marco Popillio Laenate sua legi decem milibus aeris est damnatus, quod mille jugerum agri cum filio possideret, emancipandoque filium fraudem legi fecisset."
Appian, *Bell. Civ.*, 1, 8; "τὴν γῆν ἐς τοὺς οἰκείους ἐπὶ ὑποκρίσει διένεμον."
[2] Momm., I, 389.
[3] Momm., I, 389, 390.   [4] Momm., I, 389, 390.

Sec. VIII.—Agrarian Movements between 367 and 133.

The first agrarian movement after the enactment of lex Licinia took place in the year 338, after the battle of Veseris in which the Latini and their allies were completely conquered. According to Livy,[1] the several peoples engaged in this rebellion were mulcted of a part of their land which was divided among the plebeians. Each plebeian receiving an allotment in the territory of the Latini had 2 jugera assigned him, while those in Privernum received 2¾, and those in Falernian territory received 3 jugera each (p. 252). This distribution of domain lands seems to have been spontaneous on the part of the senate. But it led to grave consequences as the Latini, indignant at their being despoiled of their lands, resorted again to arms. The plebeians, moreover, were roused to the verge of rebellion by the consul Aemilius who had been alienated from the patricians by their refusing him a triumph, and now strove to ingratiate himself with the commons by making them dissatisfied with their meagre allotments. The law, however, was carried into execution, and thus showed that the senate acquiesced in and even initiated laws when they did not in any way interfere with their possession, but referred only to territory which had just been conquered.

Agrarian Law of Curius. Beyond the distribution of the *ager publicus* which formed the basis of the numerous colonies of this period and which will be considered in their proper place, the next agrarian movement was that of Curius Dentatus. At the close of the third Samnite War the people were in great distress, as agricultural pursuits had been greatly interrupted by continued warfare. Now there seemed to be a chance of remedying this. Large tracts of land had been taken from the Samnites and Sabines, and it was now at the

---

[1] Livy, VIII, 11, 12.

disposal of the Roman [1] state for purposes of colonization and division among the impoverished citizens. In the year 287,[2] a bill was introduced by Manius Curius Dentatus, the plebeian consul for this year, and hero of the third Samnite War. He proposed giving to the citizens assignments of land in the Sabine country of seven jugera[3] each. It is certain that this bill met with great opposition but we have not been informed as to the causes.[4] It is safe to conclude, however, that the question was whether assignments of land with full right of property should be made in districts which the great land-owners wished to keep open for occupation in order that they might pasture herds thereon. The senate and the nobility so bitterly opposed the plan that the plebeians despairing of success, withdrew to the Janiculum and only on account of threatening war did they consent to the proposals of Quintus Hortensius.[5] By this move the *lex Hortensia*[6] was passed and, doubtless, the *agraria lex* was enacted at the same time although

---

[1] Ihne, I, 447.

[2] I have followed Ihne and Arnold in giving this date, but there is reason for placing it later as Valerius Maximus says, IV, 3, 5: "Manius Curius cum Italia Pyrrhum regem exegisset .... decretis a senatu septenis jugeribus agri populo."

[3] "Manii Curii nota con-cio est, perniciosum intellegi civem cui septem jugera non essent satis." Pliny, *Hist. Nat.*, XVIII.; Aurelius Victor, De Viris Illus.: Septenis "jugeribus viritim dividendis, quibus qui contentus non esset, cum perniciosum intellegi civem, nota et praeclare concione Manius Curius dictitabat." The same author speaks of four jugera being given by Curius, "Quaterna dono agri jugera viritim populo dividit" Juvenal implies a distribution of two jugera; Sat. XIV, V, 161-164:

"Mox etiam fructis aetate, ac Punica passis
Proelia vel Pyrrhum inmanem glaeosque Molossos,
Tandem pro multis vix jugera bina dabantur
Vulneribus Merces ea sanguinis atque labores."

[4] Appian, III, 5: Zonarius, VIII, 2.

[5] Ihne, I, 447.

[6] Gellius, XV, 27: "Postea lex Hortensia late, qua cautum est, ut plebisipa universum populum tenerent." Marquardt u. Momm., *Röm. Alter.*, IV, 102.

nothing definite is known concerning this point. The people must have been pacified by some other means than the mere granting of more political power. Nothing less than a share of the conquered territory would have satisfied them or induced them to return and again take up the burden of war.

Lex Flaminia. Fifty four years after the enactment of the law of Curius Dentatus, in the year 232, the tribune Caius Flaminius,[1] the man who afterwards was consul and fell in the bloody battle of lake Trasimenus, brought forward and carried a law for the distribution of the *Gallicus Ager*[2] among the plebeians. This territory[3] had been taken from the Galli Semnones fifty-one years before and was now occupied as pasture land by some large Roman families. This territory lay north of Picenum and extended as far as Ariminum[4] (Rimini.) This was an excellent opportunity for awarding lands to Roman veterans for military service, and thus to establish a large number of small farms, rather than to leave the land in the possession of the rich who resided in Rome and, consequently, formed no frontier protection against the inroads of barbarians from the north. By alloting the land, the Latin race and Latin tongue would help to Romanize territory already conquered by Roman arms. The only thing opposed to this was the possession of the land by the aristocracy. But they had no legal claim to the land and could be dispossessed without any indemnification. The senate opposed this measure to the utmost of their ability and, after all other means had failed, threatened to send an army against the tribune if he urged his bill through the tribes. They further induced his father to make use of his *potestas* in restraining his son.[5] When Flaminius was bringing up the bill for decision he was arrested by his father. "Come down, I bid thee," said the father. And the son humbled "by private

---

[1] Polyb., II, 21, 8.   [2] Varro, De R. R., 1, 2; De L. L., VI, 5.
[3] Ihne, IV, 26. See Long, I, 157, who disputes this statement.
[4] Varro, De R. R., 1, 2.; De L. L., VI, 5.   [5] Val. Max., V, 4, 5.

authority,"[1] obeyed. It finally became necessary for the plebeians to take their stand on the formal constitutional law and to cause the *agraria lex* to be passed by a vote of the assembly of the tribes without a previous resolution or subsequent approbation of the senate.[2] Polybius dates a change for the worse in the Roman constitution from this time.[3] The relief of the plebeians was further promoted by the foundation[4] of new colonies.

In the year 200, after Scipio returned as conqueror of Carthage, the senate decreed that he should be assigned some lands for his soldiers, but Livy does not tell us where they were to be assigned; whether they were to be a part of the ancient *ager publicus* or of the territory of Carthage, Sicily, or Campania, *i. e.* the new conquests of Rome. He merely says that for each year of service in Spain or Africa the soldiers were to receive two jugera each, and that[5] the distributions should be made by the *decemvirs*. In spite of the insufficiency of these details the passage reveals to us two important facts:

1. Decemvirs as well as triumvirs were at times appointed to make distributions of domain lands in accordance with the provisions of an agrarian law.

2. It reveals the profound modifications which Roman customs had passed through. The riches which began at this time to flow into Rome by reason of the many successful wars revolutionized the economic conditions of the city. It is not necessary to see only a proof of corruption in this tendency of all classes to grasp for riches and to desire luxury and ease. We must also consider that comfort was

---

[1] Val. Max., V, 4, 5; Cicero, *De Juventute*, II, 17.
[2] Ihne, IV, 26; Cicero, *De Senectute*, 4.
[3] Polybius, II, 21. [4] Livy, Epit., XX, 19.
[5] " De agris militum ejus decretum, ut quod quisque eorum annos in Hispania aut in Africa militasset, in singulos annos bina jugera acciperet, cum agrum decemviri assignarent."
Livy, XXXI, 49.

more accessible and that the price of everything, especially of the necessaries of life, had increased. In consequence of this it was difficult for soldiers to support themselves with their pay. The presents of a few sesterces given them as prize money in no way made sufficient recompense for all the miseries and privations which they had passed through during their long absence. Grants of land were the only means of recompensing their military services. This is the first example that we have found of soldiers being thus rewarded, and it consequently initiated a custom which became most frequent especially in the time of the empire. Upon the conquest of Italy which followed the expedition of Pyrrhus, the Romans found themselves led into a long series of foreign wars; Sicily furnished the stepping-stone to Africa; Africa to Spain; all these countries becoming Roman provinces. As soon as the second Punic war closed, Hannibal formed an alliance with the king of Macedonia. A war-cloud rose[1] in the east. The Ætolians asked aid from Rome, and statesmen could foretell that it would be impossible for Roman armies not to interfere between Greece and Macedonia. But these countries had been from ancient times most intimately connected with the orient, i. e., Asia, where the Seleucidae still ruled, so that a war with Greece, which was inevitable, could not fail to bring on a war with the successors of Alexander, and, these hostilities once engaged in, who could say where these accidents of war would cease, or when Roman arms could be laid aside? In this critical condition it was prudent to attach the soldiers to the republic by bonds and interests the most intimate, to make them proprietors and to assure subsistence to their families during their long absence. These wars did not much resemble those of the early republic which had for a theatre of war the country in the immediate vicinity of Rome.

---

[1] Momm., II, 230–241.

The senate continued to take the initiative in agrarian movements. In 172, after the close of the wars against the Ligurians and Gauls, we again see the senate spontaneously decreeing a new division of the lands. A part of the territory of Liguria and Cisalpine Gaul was confiscated and a *senatus consultum* ordered a distribution of this land to the commons. The praetor of the city A. Atilius, was authorized to appoint *decemvirs*, whose names Livy gives, to assign ten jugera to Roman citizens and three jugera to Latin[1] allies. Thus the senate, with a newly-born sagacity, rendered useless the demands of the tribune and recognized the justice and the utility of the agrarian laws against which it had so long protested. Indeed, it justified the propositions of the first author of an agrarian law by admitting to a share in the conquered lands the Latin allies who had so often contributed to their growth. This is the last agrarian law which Livy mentions. The Persian war broke out in this year, and an account of it fills the remaining books of this author which have come down to us. However, prior to the proposition of Tiberius Gracchus, we find in Varro[2] the mention of a new assignment of land of seven jugera *viritim*, made by a tribune named Licinius in the year 144; but the author has given such a meagre mention of it that we are unable to determine where these lands were located. If we join to these facts the cession of public territories to the creditors of the state, in 200, we shall have mentioned all agrarian laws and distributions of territory which took place before the *lex Sempronia Tiberiana* in 133.

Condition of the Country at the time of the Gracchan Rogations. During the period between 367 and 133 we find no

---

[1] Livy, XLII, 4: "Eodem anno, quum agri Ligustini et Gallici, quod bello captum erat, aliquantum vacaret, senatus-consultum factum ut is ager viritim ex senatus consulto creavit A. Atilius praetor urbanus. . . . Diverserunt dena jugera in singulos, sociis nominis Latini terna.

[2] Ihne, IV, 370.

record of serious disputes between the patricians and commons. Indeed, the senate usually took the lead in popular measures; lands were assigned without any demand on the part of the plebeians. We must not be deceived by this seeming harmony. In the midst of this apparent calm a radical change was taking place in Roman society. It is necessary for us to understand this new condition of affairs in the republic before it will be possible to comprehend the rogations of the Gracchi.

One of the greatest dangers to the republic at this time reveals itself in the claims[1] of the Italians. These people had poured out their blood for Rome; they had contributed more than the Romans themselves to the accomplishing of those rapid conquests which, after the subjugation of Italy, quickly extended the power of Rome. In what way had they been rewarded? After the terrible devastations which afflicted Italy in the Hannibalic war had ceased, the Italian allies found themselves ruined. Whilst Latium, which contained the principal part of the old tribes of citizens, had suffered comparatively little, a large portion of Samnium, Apulia, Campania, and more particularly of Lucania and Bruttium, was almost depopulated; and the Romans in punishing the unfaithful "allies" had acted with ruthless cruelty.[1] When at length peace was concluded, large districts were uncultivated and uninhabited. This territory, being either confiscated from the allies for taking part with Hannibal, or deserted by the colonists, swelled the *ager publicus* of Rome, and was either given to veterans[2] or occupied by Roman capitalists, thus increasing the revenues of a few nobles.

If a nation is in a healthful condition politically and economically so that the restorative vigor of nature is not impeded by bad restrictive laws, the devastations of land and losses of human life are quickly repaired. We might the

---

[1] Livy, XXXI. 4, 1; Ihne, IV, 370-372.   [2] Livy, *loc. cit.*

more especially have expected this in a climate so genial and on a soil so fertile as that of Italy. But Roman laws so restricted the right of buying and selling land that in every Italian community none but members of that community, or Roman citizens, could[1] buy or inherit. This restriction upon free competition, by giving the advantage to Roman citizens, was in itself sufficient to ruin the prosperity of every Italian town. This law operated continually and unobservedly and resulted in placing,[2] year by year, a still larger quantity of the soil of Italy in the hands of the Roman aristocracy. In order to palliate the evils of conquest or at least to hide their conditions of servitude, the Romans had accorded to a part of the Italians the title of allies, and to others the privileges of *municipia*.[3] These privileges were combined in a very skillful manner in the interest of Rome, but this skill did not hinder the people from perceiving that they depended upon the mere wish of the conquerors and consequently were not rights, but merely favors to be revoked at will. The Latini, who had been the first people conquered by Rome and who had almost always remained faithful, enjoyed under the name of *jus Latii* considerable privileges. They held in great[4] part the civil and political rights of Roman citizens. They were able by special services individually to become Roman citizens and thus to obtain the full *jus Romanum*. There were other peoples who, although strangers to Latium, had been admitted, by reason of their services[5] to Rome, to participate in the benefits of the *jus Latii*. The other peoples, admitted merely to the *jus Italicum*, did not enjoy any of the civil or political rights of Roman citizens, nor any of the privileges of Latin[6] allies; at best they kept some souvenirs

---

[1] Ihne, IV, 148.   [2] Ihne, IV, 371.
[3] Ihne, IV, 354; Momm., 111, 277.
[4] Momm., I, 151–162; Ihne, IV, 179. Marquardt u. Momm., IV, 26–27, 63.   [5] Livy, IX, 43, 23; Ihne, IV, 181.
[6] Ihne, IV, 185–186. Marquardt u. Momm., 46, 60.

of their departed independence in their interior administration, but otherwise were considered as subjects of Rome. And yet it was for the aggrandizement of this city that they shed their blood upon all the fields of battle which it pleased Rome to choose; it was for the glory and extension of the Roman power that they gained these conquests in which they had no share. Some who had attempted to regain their independence were not even accorded the humble privileges of the other people of Italy, but were reduced to the state of prefectures. These were treated as provinces and governed by prefects or proconsuls sent[1] out from Rome. Such were Capua, Bruttium, Lucania, the greater part of Samnium, and Cisalpine Gaul, which country, indeed, was not even considered as a part of Italy. Those who had submitted without resistance to the domination of the Romans, and had rendered some services to them, had bestowed upon them the title of *municipia*.[2] These *municipia* governed themselves and were divided into two classes:

(1.) *Municipia sine suffragio*, for example, Caere and Etruria, had only interior privileges; their inhabitants could not vote at Rome and, consequently, could not[3] participate in the exercise of sovereignty.

(2.) *Municipia cum suffragio* had, outside of their political and civil rights, the important right of voting[4] at Rome. These citizens of villages had then, as Cicero said of the citizens of Arpinum, two countries, one *ex natura*, the other *ex jure*. Lastly, there were some cities in the south of Italy, *i. e.* in Magna Graecia, that had received[5] the name of federated cities. They did not appear to be subject to Rome; their contingents of men and money were looked upon as voluntary[6] gifts; but, in reality, they were under the domination of Rome, and had, at Rome, defenders or patrons chosen because

---

[1] Marquardt u. Momm., IV, 41–43.
[2] Ibid, IV, 26.
[3] Marquardt u. Momm., IV, 27–34.
[4] Ibid.
[5] Marquardt u. Momm., IV, 44.
[6] Marquardt u. Momm., IV, 45–46.

of their influence with the Roman citizens and charged with maintaining their interests. Such was the system adopted by Rome. It would have been easy for a person in the compass of a few miles to find villages having the *jus Latii*, others with simply the *jus Italicum*, colonies, prefectures, municipia *cum et sine suffragio*. The object of the Romans was evident. They planned to govern. Cities alike in interests and patriotic motives were separated by this diversity of rights and the jealousies and hatreds which resulted from it. Concord, which was necessary to any united and general insurrection, was rendered impossible between towns, some of which were objects of envy, others, of pity. Their condition, moreover, was such that all, even the most fortunate, had something to gain by showing themselves faithful; and all, even the most wretched, had something to fear if they did not prove tractable. These Italians, with all the varied privileges and burdens enumerated above, far outnumbered the Roman citizens.[1] A comparison of the numbers of the census of 115 and that of 70 shows that the numbers of Italians and Romans were[2] as three to two. All these Italians aspired to Roman citizenship, to enjoy the right to vote to which some of their number had been admitted, and the struggle which was sometime to end in their complete emancipation had already commenced. During the first centuries of Roman history, Rome was divided into two classes, patricians and plebeians. The plebeians by heroic efforts had broken down the barriers that separated them from the patricians. The privilege of intermarriage, the possibility of obtaining the highest offices or the state, the substitution of the *comitia tributa* for the other two assemblies, had not made of Rome "an unbridled democracy," but all these benefits obtained by tribunician agitation, all the far-reaching advances gained by force of laws and not

---

[1] Momm., *Röm. Ge.*, II, 225. [2] Ihne, IV, 370.

of arms, had constituted at Rome a single people and created
a true Roman nation. There were now at Rome only rich
and poor, nobles and proletariat. With intelligence and
ability a plebeian could aspire to the magistracies and thence
to the senate. Why should not the Italians be allowed the
same privilege? It was neither just nor equitable nor even
prudent to exclude them from an equality of rights and the
common exercise of civil [1] and political liberty. The Gracchi
were the first to comprehend the changed state of affairs and
the result of Roman conquest and administration in Italy.
Their demands in favor of the Italians were profoundly
politic. The Italians would have demanded, with arms in
their hands, that which the Gracchi asked for them, had not this
attempt been made. They failed; Fulvius [2] Flaccus, Marius,[3]
and Livius Drusus [4] failed in the same attempt, being opposed
both by the nobility and the plebs.

The agrarian laws, as we have seen, had been proposed by
the senate, in the period which we are considering. How was
it then that the Gracchi had been compelled to take the initiative and that the senate had opposed them? This contradic-
tion is more apparent than real. It explains itself in great
part by the following considerations. Upon the breaking
down of the aristocracy of birth, the patriciate, the senate
was made accessible to the plebeians who had filled the
curule magistracies and were possessed of 800,000 sesterces.
Knights were also eligible to the senate to fill vacancies, and
it was this fact which caused the equestrian order to be called
seminarium senatus. For some time the new nobles, in order
to strengthen their victory and make it permanent, had formed
an alliance with the plebeians. For this reason were made
the concessions and distributions of land which the old senators were unable to hinder. These concessions were the work

---

[1] Momm., Lange, Ihne, Long—as given. [2] Momm., III, 132.
[3] Momm., III, 252, 422. [4] Momm., III, 281.

of the plebeians who had been admitted to the senate. But when their position was assured and it was no longer necessary for them to make concessions to the commons in order to sustain themselves, they manifested the same passions that the patricians had shown before them. Livy has expressed the situation very clearly: "These noble plebeians had been initiated into the same mysteries, and despised the people as soon as they themselves ceased to be despised by the patricians."[1] Thus, then, the unity and fusion which had been established by the tribunician laws disappeared and there again existed two peoples, the rich and the poor.

If we examine into the elements of these two distinct populations, separated by the pride of wealth and the misery and degradation of poverty, we shall understand this. The new nobility was made up partially of the descendants of the ancient patrician *gentes* who had adapted themselves to the modifications and transformations in society. Of these persons, some had adopted the ideas of reform; they had flattered the lower classes in order to obtain power; they profited by their consulships and their prefectures to increase or at least conserve their fortunes. Others having business capacity gave themselves up to gathering riches; to usurious speculations which at this time held chief place among the Romans. Even Cato was a usurer and recommended usury as a means of acquiring wealth. Or they engaged in vast speculations in land, commerce, and slaves, as Crassus did a little later. The first mentioned class was the least numerous. To those nobles who gave their attention to money-getting must be added those plebeians who elevated themselves from the masses by means[2] of the curule magistracies. These were insolent and purse-proud, and greedy to increase their wealth by any means in their power. Next to these two divisions of the nobility came those whom the patricians had been

---

[1] Livy, XXII, 34.   [2] Ihne, IV, 354-356.

wont to despise and to relegate to the very lowest rank under the name of *aerarii*; merchants,[1] manufacturers, bankers, and farmers of the revenues. These men were powerful by reason of their union and community of interests, and money which they commanded. They formed a third order and even became so powerful as to control the senate and, at times, the whole republic. In the time of the Punic wars the senate had been obliged to let go unpunished the crimes committed by the publican Posthumius and the means which he had employed in order to enrich himself at the expense of the republic, because it was imprudent to offend [2] the order of publicans. Thus constituted an order or guild, they held it in their hands at will to advance or to withhold the money for carrying on wars or sustaining the public credit. In this way they were the masters of the state. They also grasped the public lands, as they were able to command such wealth that no individual could compete with them. They thus became the only farmers of the domain lands, and they did not hesitate to cease paying all tax on these. Who was able to demand these rents from them? The senate? But they either composed the senate or controlled it. The magistrates? There was no magistracy but that of wealth. The tribunes and the people? These they had disarmed by frequent grants of land of two to seven jugera each, and by the establishment of numerous colonies. This was beyond doubt the real reason for their frequent distributions. They had all been made from land recently conquered. The ancient *ager* had not been touched, and little by little the Licinian law had fallen into disuetude.

---

[1] Ihne, IV, 354-356.
[2] Livy, XXV, 3: "Patres ordinem publicanorum in tali tempore offensum nolebant."

Extension of Territory by Conquest between 367 and 133.

1. Caere submitted in 353, yielding all southern Etruria to Rome.
2. Volcian territory and all Latium fell to Rome at the close of the Latin war in 339.
3. Capua, taken in 337.
4. Cales, taken in 334. In this struggle all Campania became Roman territory.
5. Sabine territory submitted in 290.
6. Tarentum, captured in 272.
7. Rhegium, captured in 270.
8. The Galli Senones were destroyed in 283 and their whole territory (Umbria) was confiscated.
9. In 293, Liguria and Transpadana Gallia were added to the Roman confederation.
10. In 222, Italy was extended to its natural boundary, the Alps, by the subjugation of the Gauls north of the Po. Of the entire territory of Italy, 93,640 square miles, fully one-third belonged to Rome. Thus, in the 287 years of the Republic, Roman territory had expanded from 115, to 31,200 square[1] miles.

At the close of the war with Hannibal, Rome further added to her territory by the confiscation of the greater part of the Gallic territory, Campania, Samnium, Apulia, Lucania, and Bruttii.

---

[1] I have not here added Roman conquests outside of the peninsula of Italy, as these conquests were not treated as Roman territory until nearly a century later.

Colonies Founded between 367 and 133.

(a). CIVIC COLONIES.

COLONIES.	PLACE.	DATES.	NO. OF C.	SIZE OF ALLOT.	JUGERA.	ACRES.
Antium.	Latium.	338	300	2	600	375
Anxur.	"	329	300	2	600	375
Minturnae.	Campania.	296	300	2	600	375
Sinuessa.	"	296	300	2	600	375
Sena Gallica.	Umbria.	283	300	6	1,800	1,125
Castrum Novum.	Picenum.	283	300	6	1,800	1,125
Aesium.	Umbria.	247	300	6	1,800	1,125
Alsium.	Etruria.	247	300	6	1,800	1,125
Fregenae.	"	245	300	6	1,800	1,125
Pyrgi.	"	191	300	6	1,800	1,125
Puteoli.	Campania.	194	300	6	1,800	1,125
Volturnum.	"	194	300	6	1,800	1,125
Liternum.	"	194	300	6	1,800	1,125
Buxentum.	Lucania.	194	300	6	1,800	1,125
Salernum.	Campania.	194	300	6	1,800	1,125
Sipontum.	"	194	300	6	1,800	1,125
Tempsa.	Bruttii.	194	300	4	1,200	750
Croton.	"	194	300	4	1,200	750
Potentia.	Picenum.	184	300	6	1,800	1,125
Pisaurum.	Umbria.	184	300	6	1,800	1,125
Parma.	Gall. Cisalp.	183	1,000	6	6,000	3,750
Mutina.	" "	183	1,000	6	6,000	3,750
Saturnia.	Etruria.	183	300	6	1,800	1,125
Graviscae.	"	181	300	5	1,500	687
Luna.	"	173	300	6	1,800	1,125
Auximum.	Picenum.	157	300	6	1,800	1,125
				Total...	38,900	30,312

(b). LATIN COLONIES.

COLONIES.	PLACE.	DATES.	NO. OF C.	SIZE OF ALLOT.	JUGERA.	ACRES.
Calles.	Campania.	334	300	4	1,200	750
Fregellae.	Latium.	328	300	4	1,200	750
Luceria.	Apulia.	314	300	4	1,200	750
Suessa.	Latium.	313	300	4	1,200	750
Pontiae.	Isle of Latium.	313	300	4	1,200	750
Saticula.	Samnium.	313	300	4	1,200	750
Sora.	Latium.	312	4,000	4	1,200	750
Alba.	"	303	6,000	6	36,000	22,500
Narnia.	Umbria.	299	300	6	1,800	1,125
Carseoli.	Sabini.	298	4,000	6	24,000	15,000
Venusia.	Apulia.	291	300	6	1,800	1,125
Hatria.	Picenum.	289	300	6	1,800	1,125
Cosa.	Campania.	273	1,000	6	6,000	3,750
Paestum.	Lucania.	273	300	6	1,800	1,125
Ariminum.	Agr. Gallicus.	268	300	6	1,800	1,125
Beneventum.	Samnium.	268	300	6	1,800	1,125
Firmum.	Picenum.	264	300	6	1,800	1,125
Aesernia.	Samnium.	263	300	6	1,800	1,125
Brundisium.	Calabria.	244	300	6	1,800	1,125
Spoletium.	Umbria.	241	300	6	1,800	1,125
Cremona.	Gaul.	218	6,000	6	36,000	22,500
Placentia.	"	218	6,000	6	36,000	22,500
Copiae.	Lucania.	193	300	6	1,800	1,125
Bononia.	Gaul.	192	3,000	6	18,000	11,250
Aquileia.	"	181	4,500	6	27,000	16,875
		Total.............			211,200	132,000
		Civic Colonies..........			38,900	30,312
		Grand Total............			250,100	162,312 or 253.61 Sq. Mi.

Sec. 9.—Latifundia.

"After having pillaged the world as praetors or consuls during time of war, the nobles again pillaged their subjects as governors in time of peace;[1] and upon their return to Rome with immense riches they employed them in changing the modest heritage of their fathers into domains vast as provinces. In villas, which they were wont to surround with forests, lakes and mountains . . . where formerly a hundred families lived at ease, a single one found itself restrained. In order to increase his park, the noble bought at a small price the farm of an old wounded soldier or peasant burdened with debt, who hastened to squander, in the taverns of Rome, the modicum of gold which he had received. Often he took the land without paying anything.[2] An ancient writer tells us of an unfortunate involved in a law suit with a rich man because the latter, discommoded by the bees of the poor man, his neighbor, had destroyed them. The poor man protested that he wished to depart and establish his swarms elsewhere, but that nowhere was he able to find a small field where he would not again have a rich man for a neighbor. The nabobs of the age, says Columella, had properties which they were unable to journey round on horseback in a day, and an inscription recently found at Viterba, shows that an aqueduct ten miles long did not traverse the lands of any new proprietors. . . . The small estate gradually disappeared from the soil of Italy, and with it the sturdy population of laborers. . . . Spurius Ligustinus, a centurian, after twenty-two campaigns, at the age of more than

---

[1] Cicero says these exactions were common and that the provinces were even restrained from complaining. Verres apologized for his exactions by saying that he simply followed the common example. In Verrem, II, 1-3, 17.

[2] "Parentes aut parvi liberi militum, ut quisque potentiori confinis erat, sedibus pellebantur." Sall., Jugertha, 41. Horace, Ode II, 18.

fifty years, did not have for himself, his wife, and eight children more than a jugerum of land and a cabin."[1]

To this masterly sketch quoted from Duruy, we can but add a few facts. Pliny affirms that under Nero only six men possessed the half of Africa.[2] Seneca, who himself possessed an immense fortune, says, concerning the rich men of his time, that they did not content themselves with possessing the lands that formerly had supported an entire people; they were wont to turn the course of rivers in order to conduct them through their possessions. They[3] desired even to embrace seas within their vast domains. We must here, it is true, make some allowance for rhetoric. So, too, in the writings of Petronius, some allowance for satire must be made, where he represents the clerk of Trimalchio making a report of that which has taken place in a single day upon one of the latter's farms near Cumae. Here on the 7th of the calends[4] of July, were born 30 boys and 40 girls; 500,000 bushels of wheat were harvested and 500 oxen were yoked. The clerk goes on to say that a fire had recently broken out in the *Gardens of Pompey*, when he is interrupted by Trimalchio asking when the *Gardens of Pompey* had been purchased for him, and is informed that they had been in his possession for a year.[5] So it appears that Trimalchio, in whom Petronius has personified the pride, the greed, and the vices of the rich men of his time, did not know that he was the possessor of a magnificent domain. In another place

---

[1] Duruy, *Hist. des Romains*, II, 46-47.
[2] "Sex domini semissem Africae possidebant." *Hist. Nat.*, XVIII, 7.
[3] Seneca, Epist., 89.
[4] Petronius, Sat., 48: VII. calendas sextilis in praedio Cumano, quod est Trimalchionis, nati sunt pueri, XXX, puellae, XL; sublata in horreum, ex area, tritici millia modium quingenta; boves domiti quingenti . . . eodem die incendium factum est in hortis Pompeianis, ortum ex aedibus nastae, villici.
[5] Quid? inquit Trimalchio: quando mihi Pompeiani horti emti sunt? Anno priore, inquit actuarius. (*Ibid.* 53.)

Petronius causes Trimalchio to say that everything which could appeal to the appetite of his companions is raised upon one of his farms which he has not yet visited and which is situated in the neighborhood of Terracina and Tarentum, towns[1] which are separated by a distance of 300 miles. Finally, led on by his immoderate desire to augment his riches and increase his possessions, the hero of Petronius asks but one thing before he dies, *i. e.*, to add Apulia[2] to his domains; he, however, admits that he would not take it amiss to join Sicily to some lands which he owned in that locality or to be able, should envy not check him, to pass into Africa[3] without departing from his own possessions. All this has a basis of fact. Trimalchio would never have been created, had not the favorite freedmen of Nero crushed the people by their luxury, debauches, and scandals.

But the condition of society pictured by Seneca and Petronius is that of the first century of the Christian era and might not be taken to represent the condition of affairs in the second century B. C., had we not some data which go to prove the concentration of property, the disparity between classes, and the depopulation of Italy within the same century as the Gracchi. Cicero was not considered one of the richest men in Rome, yet he possessed many villas, and he has himself told us that one of them cost him 3,500,000 sesterces, about $147,000.[4] Cornelia, the mother of the Gracchi, had a country residence in the vicinity of Micenum which cost[5]

---

[1] Vinum, inquit, si non placet, mutabo; vos illud, oportet faciatis. Deorum beneficio nōn emo, sed nunc, quidquid ad salivam facit, in suburbano nascitur eo quod ego adhuc non navi. Dicitur confine esse Tarracinensibus et Tarentinis.

[2] Quod si contigerit Apuliae fundos jungere, satis vivus pervenero, (*Ibid.* 77.)

[3] Nunc conjungere agellis Siciliam volo, ut quum Africam libuerit ire, per meos fines navigem. Sat., 48.

[4] Ad Fam., V, 6: "quod de Crasso domum emissem emi eam ipsam domum H. S., XXXV."

[5] Plutarch, *Life of Marius*.

75,000 drachmae ($14,000); Lucullus some years afterwards bought it for 500,200 drachmae ($100,040). According to Cicero,[1] Crassus had a fortune of 100,000,000 sesterces ($4,200,000). This does not astonish us when we see upon the *via Appia*, near the ruins of the circus of Caracalla and but a short distance from the Catacombs of St. Sebastian and the fountain of Aegeria, the still important remains of the tomb of Caecilia Metella, daughter of Metellus Creticus and wife of the tribune Crassus, as the inscription testifies. It is a vast "funereal fortress" constructed of precious marble, and which gives us the first example of the luxury afterwards so common among the Romans. Then, too, we remember that Crassus was wont to say that no one was rich who was not able to support an army with his revenues, to raise six legions and a great number of auxiliaries, both infantry and cavalry.[2]

Pliny confirms this statement concerning Crassus, but adds that Sulla was even richer.[3] Plutarch gives us fuller details and also explains the origin of the colossal fortune of Crassus. According to him Crassus had 300 talents ($345,000), with which to commence. Upon his departure for the Parthian war in which he lost his life, he made an inventory of his property and found that he was possessed of 7,100 talents, $8,165,000, double what Cicero attributes to him. How did Crassus increase his fortune so enormously? Plutarch says that he bought the property confiscated by Sulla at a very low figure. Then, he had a great number of slaves distinguished for their talents; lecturers, writers, bankers, business men, physicians, and hotel-keepers, who turned over to him the benefits which they realized in their diverse industries. Moreover, he had among his slaves 500 masons and architects. Rome was built almost

---

[1] De Repub., III, 7: Cur autem, si pecuniae modus statuendus fuit feminis, P. Crassi filia posset habere, si unica patri esset, aeris millies, salva lege?

[2] Cicero, *Paradoxia*, VI.   [3] Pliny, *Hist. Nat.*, XXXIII, 10.

entirely of wood and the houses were very high, consequently fires were frequent and destructive. As soon as a fire broke out, Crassus hastened to the place with his throng of slaves, bought the now burning buildings—as well as those threatened—at a song, and then set his slaves to work extinguishing the fires. By this means he had become possessed of a large[1] part of Rome.

Some other facts confirm that which Plutarch tells us of Crassus. Athenaeus[2] says that it was not rare to find Roman citizens possessed of 20,000 slaves. At the commencement of the civil war between Cæsar and Pompey, the future dictator found opposed to him, in Picenum, Domitius[3] Ahenobarbus at the head of thirty cohorts. Domitius seeing his troops wavering, promised to each of them four jugera out of his own possessions, and a proportionate part to the centurians and veterans. What must have been the fortune of a man who was able to distribute out of his own lands, and surely without bankrupting himself, about 100,000 jugera?

Sec. 10.—The Influence of Slavery.

The last of the evils which we wish to mention as bringing about the deplorable condition of the plebeians at the time of the Gracchi, and which brought more degradation and ruin in its train than all the others, is slavery. Licinius Stolo had attempted in vain to combat it. Twenty-four centuries of fruitless legislation since his death has scarcely yet taught the most enlightened nations that it is a waste of energy to regulate by law the greatest crime against humanity, so long as the conditions which produced it remain the same. The Roman legions, sturdy plebeians, marched on to the conquest of the world. For what? To bring home vast throngs of captives who were destined, as slaves, to eat the bread, to sap

---

[1] Plutarch, *Crassus*, c. 1 and 2.
[2] Athenaeus, *Deipnosophistae*, VI, 104.   [3] Cæsar, *Bell. Civ.*, 1, 17.

the life blood, of their conquerors. The substitution of slaves for freemen in the labors of the city and country, in the manual arts and industries, grew in proportion to the number of captives sold in the markets of Rome. All the rich men followed more or less the example of Crassus; they had among their slaves, weavers, carvers, embroiderers, painters, architects, physicians, and teachers. Suetonius tells us that Augustus wore no clothing save that manufactured by slaves in his own house. Atticus hired his slaves to the public in the capacity of copyists. Cicero used slaves as amanuenses. The government employed slaves in the subordinate posts in administration; the police, the guard of monuments and arsenals, the manufacture of arms and munitions of war, the building of navies, etc. The priests of the temples and the colleges of pontiffs had their familiae of slaves.

Thus in the city, plebeians found no employment. Competition was impossible between fathers of families and slaves who labored *en masse* in the vast work-shops of their masters, with no return save the scantiest subsistence, no families, no cares, and most of all no army service. In the country it was still worse. It would appear that none but slaves were employed in the cultivation of the land. Doubtless the number of slaves in Italy has been greatly exaggerated, but it is certain that the substitution of slave labor for free, was an old fact when Licinius[1] attempted by the formal disposition of his law to check the evil. In the first centuries of Rome, slaves must have been scarce. They were still dear in the time of Cato, and even Plutarch mentions as a proof of the avarice of the illustrious[2] censor, that he never paid more than 15,000 drachmae for a slave. After the great conquests of the Romans, in Corsica, Sardinia, Spain, Greece, and the Orient, the market went down by reason of the multitude of human

---

[1] M. Dureau de la Malle, *Ec. polit. des Romains.* ch. 15, p. 143; ch. 2, p. 231.

[2] Plutarch, *Cato the Censor*, 6 and 7.

beings thrown upon it. An able-bodied, unlettered man could be bought for the price of an ox. Such were the men of Spain, Thrace, and Sardinia. Educated slaves from Greece and the East brought a higher price. We learn from Horace, that his slave Davus whom he has rendered so celebrated, cost him 500 drachmae.[1] Diodorus of Siculus says that the rich caused their slaves to live by their own exertions. According to him the knights employed great bands of slaves in Sicily, both for agricultural purposes and for herding stock, but they furnished them with so little food that they must either starve or live by brigandage. The governors of the island did not dare to punish these slaves for fear of the powerful order which owned them.[2] Slave labor was thus adopted for economic reasons, and, for the same reasons, agriculture in Italy was abandoned for stock raising.

Says Varro:[3] "Fathers of families rather delight in circuses and theatres than in farming and grape culture. Therefore, we pay that wheat necessary for our subsistence be imported from Africa and Sardinia; we pick our grapes in the isles of Cos and Chios. In this land where our fathers who founded Rome instructed their children in agriculture, we see the descendants of those skillful cultivators, by reason of avarice and in contempt of laws, transferring arable lands into pasture fields, perhaps ignorant of the fact that agriculture and fatherland were one."

Fewer men were needed for the care of these pasture lands; but the evil did not stop here. Little by little these pasture lands were transformed into mere pleasure grounds attached to villas. This had already begun to take place as early as the second Punic war, when the plains of Sinuessa[4] and

---

[1] Horace, Sat. II, 7; v. 42-43: "Quid? si me stultior ipso quingentis empto drachmis, deprehenderis."
[2] Diodorus, Siculus, Fg. of Bk. XXXIV.
[3] Varro, De R. R. Proem. 3, 4.   [4] Livy, XXII, 15.

Falernia were cultivated rather for pleasure than the necessaries of life; so that the army of Fabius could find nothing upon which to sustain itself. Under these influences the plebeians, in 133, had become merely a turbulent, restless mass, but full of the activity and the energy which had characterized them in the early centuries of the republic. They were composed chiefly of the descendants of the ancient plebeian families, decimated by wars and by misery. They were the heirs of those for whom Spurius Cassius, Terentillius Arsa, Virginius, Licinius Stolo, Publilius Philo, and Hortensius had endured so many conflicts and even shed their blood; but they had become brutalized by poverty, debauchery, and crime. No longer able to support themselves by labor, they had become beggars and vagabonds.

SEC. 11.—LEX SEMPRONIA TIBERIANA.

In 133, more than two centuries after the enactment of the law of Licinius Stolo, Tiberius Gracchus, tribune of the people for that year, brought forward a bill which was in fact little less than a renewal of the old law. It provided that no one should occupy more than five hundred jugera of the *ager publicus*, with the proviso that any father could reserve[1] 250 jugera for each son.[2] This law differed from that of Licinius in that it guaranteed permanent possession

---

[1] App., I, 9; Livy, Epit., LVIII, XII: "possessores, qui filios in potestate haberent, supra legitimum modum ducena quinquagena jugera in singulos retinerent."

[2] Mommsen states that this privilege was limited to 1000 jugera in all, and Wordsworth follows him, making the same statement. Lange, *Röm. Alterthümer*, III, 9, agrees with Mommsen and cites, App. B. C., I, 9, 11; Vell., 2, 6; Livy, Ep., 58; Aurelius Victor, 64; Sic. Flacc., p. 136, Lach. I find no direct proof in the places mentioned of what Lange asserts while App. (I, 11), says: "καὶ παισὶ, οἷς εἰσὶ παῖδες ἑκάστῳ καὶ τούτων τὰ ἡμίσεα." Long says there is no proof of any limitation as to number of sons, while Ihne, Duruy and Nitzsch are agreed in following the statement of Appian, as I have here done. See Marquardt u. Momm., *Röm. Alter*, 106.

of this amount to the occupier and his heirs forever.¹ Other clauses were subjoined providing for the payment² of some equivalent to the rich for the improvements and the buildings upon the surrendered estates, and ordering the division of the domain thus surrendered among the poorer citizens in lots of 30 jugera each, on the condition that their portions should be inalienable.³ They bound themselves to use the land for agricultural purposes and to pay a moderate rent to the state. It appears that the Italians were not excluded from the benefit of this law.⁴

The design of this bill was to recruit the ranks of the Romans by drafts of freeholders from among the Latins. Such as had been reduced to poverty were to be restored to independence. Such as had been sunk beneath oppression were to be lifted up to liberty.⁵ No more generous scheme had ever been brought before the Romans. None ever met with more determined opposition, and for this there was much reason. There might have been some like the tribune's friends ready to part with the lands bequeathed to them by their fathers; but where one was willing to confess, a hundred stood ready to deny the claim upon them. Nor had they any such demands to meet as those of the olden times. Then the plebeians were a firm and compact body which

---

[1] App., I, 11.
[2] Momm., III, 114; Plutarch, *Tiberius Gracchus*, 9, l. 9.
[3] App., I, l. 3.
[4] App., I, 9: "Τιβέριος Γράκχος . . . δημαρχῶν ἐσεμνολόγησε περὶ τοῦ Ἰταλικοῦ γένους ὡς εὐπολεμωτάτου τε καὶ συγγενοῦς, φθειρομένου δὲ κατ' ὀλίγον ἐς ἀπορίαν καὶ ὀλιγανδρίαν. Also App. B. C., I, 13; Γράκχος δὲ μεγαλαυχούμενος ἐπὶ τῷ νόμῳ . . . οἷα δὴ κτίστης οὐ μιᾶς πόλεως οὐδ' ἑνὸς γένους ἀλλὰ πάντων ὅσα ἐν Ἰταλίᾳ ἔθνη, ἐς τὴν οἰκίαν παρεπέμπετο."
Ihne, IV, 385. Lange says (III, 10): "Das Gracchus die Latiner und Bundesgenossen nicht berücksichtigte, war bei der Gesinnung der römischen Bürgerschaft gegen die Latiner ganz natürlich." I can not see how he harmonizes this statement with that of App., Ἰταλικοῦ γένους and Ἰταλίᾳ ἔθνη. Momm., *Röm. Ge.*, II, 88.
[5] Sallust, *Jugertha*, XLII.

demanded a share of recent conquests that their own blood and courage had gained. Now it was a loose and feeble body of various members waiting for a share in land long since conquered, while their patron rather than their leader exerted himself for them.

Tiberius, like Licinius, met with violent opposition, but he had not like him the patience and the fortitude to wait the slower but safer process of legitimate agitation. He adopted a course[1] which is always dangerous and especially so in great political movements. Satisfied with the justice of his bill and stung by taunts and incensed by opposition, he resolved to carry it by open violation of law. He caused his colleague, Octavius, who had interposed his veto, to be removed from office by a vote of the citizens—a thing unheard of and, according to the Roman constitution, impossible—and in this way his bill for the division of the public land was carried and became a law. It required the appointing of three commissioners to receive and apportion the public domain.[2] This collegium of three persons,[3] who were regarded as ordinary and standing magistrates of the state, and were annually elected by the assembly of the people, was entrusted with the work of resumption and distribution. The important and difficult task of legally settling what was domain land and what was private property was afterward added to these functions. Tiberius himself, his brother Caius, then at Numantia, and his father-in-law, Claudius, were nominated, according to the usual custom of intrusting the execution of a law to its author and his chosen

---

[1] App., I, XII; Plutarch, *Tiberius Gracchus*, X-XII; Julii Flori Epitoma, II, (Biblioth. Teubner, p. 67): "Sit ubi intercedentem legibus suis C. Octavium vidit Gracchus, contra fas collegii, juris, potestas, is injecta manu depulit rostris, adeoque praesenti metu mortis exterruit, ut abdicare se magistratu cogeretur."

[2] Momm., III, 115.

[3] App., I, 9; Livy, *Epit.*, LVIII, 12; Plut., *Tib. Gr.*, 8-14; Cic., De Leg. Agr., II, 12, 13; Velleius, 2, 2; Aurelius Vic., De Vir. Illus., 64.

adherents.[1] The distribution was designed to go on continually and to embrace the whole class that should be in need of aid. The new features of this agraria lex of Sempronius, as compared with the Licinio-Sextian, were, first, the clause in favor of the hereditary possessors; secondly, the payment of quit-rent, and inalienable tenure proposed for the new allotments; thirdly, and especially, the permanent executive, the want of which, under the older law, had been the chief reason why it had remained without lasting practical application.[2]

The dissatisfaction of the supporters of the law concurred with the resistance of its opponents in preventing its execution or at least greatly embarrassing the collegium. The senate refused to grant the customary outfit to which the commissioners[3] were entitled. They proceeded without it. Then the landowners denied that they occupied any of the public land, or else asked such enormous indemnities as to render the recovery impossible without violence. This roused opposition. The *ager publicus* had never been surveyed, private boundaries had in many cases been obliterated, and, except where natural boundaries marked the limit of the domain land, it was impossible to ascertain what was *ager publicus* and what *ager privatus*. To avoid this difficulty the commission adopted the just but hazardous expediency of throwing the burden of proof upon the occupier. He was summoned before their tribunal and, unless he could establish his boundaries or prove that the land in question had never been a part of the domain land, it was declared *ager publicus* and confiscated.[4]

On the other hand the newly made proprietors were contending with one another, if not with the commissioners. The

---

[1] Plutarch, *Tiberius Gracchus*, 13.
[2] Momm., III, 115. See Ihne's just condemnation of this clause; IV, 387.
[3] Plutarch, *Tib. Grac.*, XIII, ln. 12; Duruy, *Hist. Rom.*, vol. II, pp. 339-420 of Translation.
[4] Long, I, 183; Ihne; IV. 387; Lange, III, 10-12; Nitzsch, *Die Gracchen*, 294 *et seq*.

Italians were, in some cases, despoiled instead of relieved by the law. The complaints of those turned out of their estates to make room for the clamorous swarms from the city, drowned the thanks of such as obtained a portion of the lands. Not even with the wealth of Attalus had Tiberius bought friends enough to aid him at this time.[1] The same spirit of lawlessness which he himself had invoked in the passing of his law, was in turn made use of by his enemies to crush him. Having been absent from Rome while performing his duties as commissioner, he now returned as a candidate for re-election to the tribunate, a thing in itself contrary to law, and in the struggle which arose over his re-election, was slain a little more than six months after his appointment[2] to membership in the collegium.

Uncertainty as to the Details of the Lex Sempronia. We are very imperfectly informed upon many points in Tiberius' agrarian law. In the first place, the question arises, were those persons holding less than 500 jugera at the time of its enactment given their lands as *bona fide* private property with the privilege of making up the deficiency? If not, then the law, instead of punishing, would seem to reward violation of its tenets, and he who had with boldness appropriated the greatest quantity of domain land would now be an object of envy to his more honest but less fortunate neighbors.

Secondly, what arrangement was made as to the buildings and improvements already upon the land? Were these handed over to the new owners without any payment on their part? This would work great inequality in the value of allotments made, and yet we cannot see where the poor man was to obtain the money to pay for these. Then again, what was to become of the numerous slaves which had hitherto carried on the agriculture now destined to be performed by

---

[1] Plutarch, *Tib. Grac.*, 14; Florus, II.
[2] Cicero, *De Amicitia*, 12. "Tiberius Gracchus regnum occupare conatus est vel regnavit is quidem paucas menses."

small holders? Their masters would have no further use for them and would consequently swell the lists of freedmen in order to avoid the expense of feeding them. This law was passed in the midst of the Sicilian slave war and Tiberius Gracchus would surely not have neglected to make some provision to meet this exigency. The law as it stands in its imperfect condition seems to be the work of an ignorant, unprincipled political charlatan, but we are convinced Tiberius was not that. Moreover, we know that he had the help of one of Rome's most able lawyers, Publius Mucius Scaevola, and the advice of his father-in-law, Appius Claudius, who was something of a statesman. We are therefore convinced that some conditions which were to meet these obstacles were enacted. We must admit, however, that it is a little surprising that no fragment of such conditions has ever reached us in the literature of Rome.

Results of this Law. Although Tiberius was dead, yet his law still lived, and, indeed, received added force from the death of its author. The senate killed Gracchus but could not annul his law. The party which was favorable to the distribution of the domain land gained control of affairs. Gaius Gracchus, Marcus Fulvius Flaccus, and Gaius Papirius Carbo, were the chief persons in carrying the law into effect. Mommsen (vol. III, p. 128) says: "The work of resuming and distributing the occupied domain land was prosecuted with zeal and energy; and, in fact, proofs to that effect are not wanting." As early as 622 the consul of that year, Publius Popillius, the same who presided over the prosecution of the adherents of Tiberius Gracchus, recorded on a public monument that he was 'the first who had turned the shepherd out of the domains and installed farmers in their stead;' and tradition otherwise affirms that the distribution extended over all Italy, and that in the formerly existing communities the number of farmers was everywhere augmented—for it was the design of the Sempronian agrarian law to elevate the former class, not by the founding of new communities, but by the strengthening of those already in existence.

"The extent and the comprehensive effect of these distributions are attested by the numerous arrangements in the Roman art of land-measuring referable to the Gracchan assignations of land; for instance, the due placing of boundary stones, so as to obviate future mistakes, appears to have been first suggested by the Gracchan courts for defining boundaries and by the distribution of land.

"But the number on the burgess-rolls gives the clearest evidence. The census, which was published in 623, and actually took place probably in the beginning of 622, yielded not more than 319,000 burgesses capable of bearing arms, whereas six years afterwards (629), in place of the previous falling off (p. 108), the number rises to 395,000, that is 76,000 of an increase beyond all doubt solely in consequence of what the allotment commission did for Roman burgesses."

Ihne says, concerning this same commission (vol. IV, p. 409): "The triumvirs entered upon their duties under the most unfavorable circumstances. . . . We may entertain serious doubts whether they or their immediate successors ever got beyond this first stage of their labors, and whether they really accomplished the task of setting up any considerable number of independent freeholders." Ihne further says (vol. IV, p. 408, n. 1), in answer to the statements made by Mommsen, which we have quoted above: "There is an obvious fallacy in this argument, for how could the assignment of allotments to poor citizens increase the number of citizens? There is nothing to justify the assumption that non-citizens were to share in the benefit of the land-law, and that by receiving allotments they were to be advanced to the rank of citizens. If the statements respecting the census of 131 B. C. and 125 B. C. are to be trusted, the great increase in the number of citizens must be explained in another way. It is possible . . . that after the revolt of Fregellae (125 B. C.) a portion of the allies were admitted to the Roman franchise by several plebiscites. We know nothing of such plebiscites; but it is not unlikely that the Roman senate in 125 B. C. acted on the principle of making timely concessions to a portion of the rebels, and thus preventing unanimous action among them. This is what was done in 90 B. C. during the great Social War. By such an admission of allies, the increase of citizens between 131 and 125 might possibly be explained."

If we examine the objections which Ihne raises we shall not find them so formidable as first appears. Mommsen does not say that the number of citizens was increased. What he does say is that the number of burgesses capable of bearing arms was increased (vol. III, p. 128). In 570–184, the Servian Military Constitution was so modified as to admit to service in the burgess army, persons possessed of but 4,000 asses ($85). In case of need all those who were bound to serve in the fleet, i. e. those rated between 4,000 and 1,500 asses and all freedmen, together with the free-born rated between 1,500 asses ($30) and 375 asses ($7.50), were enrolled in the burgess infantry.[1] It is easy enough to see that the gift on the part of the government of 30 jugera (24 acres) of land to each poor citizen, would raise him from the ranks of the proletariate and make him liable to military service.

This is sufficient to establish Mommsen's thesis;[2] and it is not necessary to consider the second point, viz., that non-citizens were not to share in the benefit of the land law nor thereby to be raised to the rank of citizens, although to us it would be no more difficult to believe this than that 76,000 allies had been admitted to the Roman franchise "by several plebiscites" no trace or rumor of which had been preserved.

It can hardly be supposed that the Italian farmers were multiplied at the same ratio as were the Romans; but the result must have been most beneficial even to them.

---

[1] Momm., II, p. 417.

[2] Professor Long thinks that the law of Tiberius soon became a dead letter. Lange (*Röm. Alter.*, III, 26–29), inclines to this view. Duruy (II, 419–420), and most other modern writers agree with Mommsen.

In the accomplishing of this result, respectable interests and existing rights were no doubt violated. The commission itself was composed of violent partisans who, being judges unto themselves, did not scruple to carry out their plans even at the cost of recklessness and tumult. Loud complaints were made, but usually to no avail. If the domain question was to be settled at all, the matter could not be carried through without some such rigor of action. Intelligent Romans wished to see the plan thoroughly tested. But this acquiescence had a limit. The Italian domain was not all in the hands of Roman citizens. Allied communities held the usufruct of large tracts of it by means of decrees of the people or the senate, and other portions had been taken possession of by Latin burgesses. These in turn were attacked by the commissioners; but to give fresh offense to these Latini, who were already overburdened with military service, without share in the spoils, was a matter of doubtful policy.

The Latini appealed to Scipio in person, and by his influence a bill was passed by the people which withdrew from the commission its jurisdiction and remitted to the consuls the decision as to what were private and what domain lands. This was a mild way of killing the law, and resulted in that. It had, however, in great measure, fulfilled its object and left little territory in the hands of the Roman state.

SEC. 12.—LEX SEMPRONIA GAIANA.

Gaius Gracchus really enacted no new agrarian law but merely re-established the power of the commission which had been appointed by his brother ten years before; which power they had lost by the law of Scipio.[1] Gaius' law was enacted merely to preserve the principle, and the distribution of land,

---

[1] Scipio must have caused a plebiscitum to be enacted, for the repeal of this clause, as an existing law could not be repealed by a *senatus consultum*. See Ihne, IV, 414, note.

if resumed at all, was on a very limited scale. This is made known from the fact that the burgess-roll showed precisely the same number capable of bearing arms in 124 and 114. As has already been stated, the domain land had been exhausted by the commission before losing its power, and, therefore, Gaius had none to distribute.[1] The land held by the Latini could only be taken into consideration with the difficult question of the Roman franchise. But when Gaius proposed the establishment of colonies in Italy, at Tarentum and Capua, whose territories had been hitherto reserved as a source of revenue to the treasury,[2] he went a step beyond his brother and made this also liable to be parcelled out; not, however, according to the method of Tiberius, who did not contemplate the establishment of new communities, but according to the colonial system. There can be little doubt that Gaius designed to aid in permanently establishing[3] the revolution by means of these new colonies in the most fertile part of all Italy. His overthrow and death put a stop to the establishment of the contemplated colonies and left this territory still tributary to the treasury.

---

[1] Momm., III, 137.

[2] Cicero, *De Leg. Agr.*, II, c. 29–32; Marquardt u. Momm., *Röm. Alter.*, IV, 106: "ager publicus mit Ausnahme einiger dem Staate unenbehrlicher Domainen, wozu namentlich das Gebiet von Capua und das stellatische Feld bei Cales gehörte."

[3] Ihne, IV, 438–479. Plutarch, *Gaius Gracchus*, 13.

CHAPTER III.

SEC. 13.—LEX THORIA.[1]

According to Appian, during the years which followed the death of Gaius Gracchus up to the tribunate of Saturninus, that is to say, between the years 120 and 100, three agrarian laws were proposed and adopted.

1. A law "That the holders of the land which was the matter in dispute might legally sell [2] it." Appian, who is the only authority for this period, does not give the date of the law nor the name of the tribune who proposed it, but Ihne [3] makes the date 118, and Mommsen assigns the law to Marcus [4] Drusus. This law was a repeal of all the restrictions which the Gracchi had placed upon assignments of public land. The object of this clause was to secure the success of their great reforms, and to establish a number of small proprietors who would cultivate their little farms, and breed citizens and soldiers. But forced cultivation is impossible, and sumptuary laws have never yet succeeded in increasing [5] population. Again it is inconsistent to give land to a man and deprive him of the power of sale, for this is an essential part of that domain which we call property in land. If a man wishes to sell, he

---

[1] Rudorff, *Ackergesetz des Spurius Thorius*, Zeitschrift für geschichtliche Rechtswissenschaft, Band X, s. 1-158. Corpus Inscriptionum Latinarum, vol. V, pp. 75-86. Wordsworth, *Specimens and Fragments of Early Latin*, 440-459.
[2] Appian, *Bell. Civ.*, I, c. 27.   [3] Ihne, *Roman History*, V, 9.
[4] Momm., *Rom. Hist.*, III, 165.
[5] Long, *Decline of the Rom. Rep.*, I, 352. See Lange, *Röm. Alter.*, III, 48.

will always have sufficient reasons for so doing, and a rich man can afford to pay[1] the highest price, freedom of exchange thus bringing ultimate good to both parties. It is easy to comprehend the consequences of this law. It was the commencement of a reaction entirely aristocratic in its nature.[2] It was skillfully conducted with the ordinary spirit of the Roman senate, the ruses, mental reservations, and dissimulations under guise of public interest. The aristocracy presented to the plebeian farmers, established by the lex Sempronia, a means of promptly and easily satisfying their passions. They had never earned their little farms, nor did they appreciate the independence of the tiller of the soil. Unaccustomed to farm labor,[3] and the plodding unexciting life of the Roman *agricola*, they made haste to abandon a toilsome husbandry, the results of which seemed to them slow and uncertain, and with the pieces of silver which they received as the price of their lands, returned to Rome to swell the idle and vicious throng[4] which enjoyed the sweet privilege of an existence sustained without labor.

Thus the nobles re-entered promptly and cheaply into the possession of the lands of which Tiberius had but a short time before deprived them, and, by means of a little sacrifice, substantially and legally converted their possessions into real property, while the plebeians whom Tiberius had wished to elevate by means of forcing[5] upon them the necessity of labor, fell back into their accustomed poverty and brutality. But the object for which the nobles were striving was not yet completely gained. The present victory was theirs; they now strove to guarantee the future, and so render impossible dangers similar to those already passed through.

2. A second law was thus enacted: "Spurius Borius, a tribune, proposed a law to this effect; that there should be

---

[1] Long, *loc. cit.*   [2] Momm., III, 161; Ihne, V, 10.
[3] Long, *loc. cit.*   [4] Lange, III, 48–49; Marquardt u. Momm., IV, 108.
[5] Long, *loc. cit.*   Momm., III, 167–168; Ihne, V, 8–10.

no more distribution of the public land, but it should be left to the possessors who should pay certain charges (*vectigalia*) for it to the state (δήμῳ) and that the money arising from these payments should be distributed."[1]

It is easy to comprehend the effect of a law so conceived. On the one hand it guaranteed to the possessors full property in the public lands which they held. From this point of view it was aristocratic. But on the other hand it aimed to unite the interests of the common people with those of the aristocracy, by placing a tax of one tenth of the produce upon the holders of these lands,[2] thus reëstablishing the law which had been annulled by Drusus. This took the place of distributions of land, which had now been made impossible[3] in Italy. In reality this law was disastrous to the plebeians as it established a tax[4] for their benefit, a *congiarium*, and placed a premium upon laziness.

The narration of Appian presents some grave difficulties. In all the manuscripts of Appian the name of the tribune proposing the second law is Spurius Borius.[5] Cicero mentions a tribune by the name of Spurius[6] Thorius and Schweighäuser in his edition of Appian has changed 'Borius' to 'Thorius.' But this does not lessen the difficulty, as the law which Cicero attributes to Thorius is entirely different from the second law of Appian which, according to him was introduced by Spurius Borius. Cicero says that Spurius Thorius "freed the public lands from the vectigal."[7] Appian says that Spurius Borius guaranteed the *possessions* in the public lands, levying a tax on them for the benefit of the people. It is a sheer waste of time to attempt to harmonize these two statements.[8] Granting that Spurius Borius and

---

[1] Appian, I, c. 27.
[2] Long, I, 353.
[3] Long, I, 354.
[4] Ihne, V, 10-11.
[5] Long, I, 353; Wordsworth, 440; Momm., III, 165, note; Ihne, V, 9; Lange, III, 48; Appian, I, c. 27.
[6] Cicero, *Brut.*, 36.
[7] Cicero, *De Orat.*, II, 70.
[8] Marquardt u. Momm., *Röm. Alter.*, IV, 108, n. 4; Wordsworth, 441.

Spurius Thorius are one and the same person, the statements still remain diametrically opposed according to a simple and commonly accepted translation of Cicero's words: "Sp. Thorius satis valuit in populari genere dicendi, is qui agrum publicum vitiosa et inutile lege vectigali levavit." Mommsen makes Cicero agree with Appian by changing "vectigali" into the instrument, and rendering[1] "relieved the public land from a vicious and useless law by imposing a vectigal." No other writer agrees with Mommsen in making such a translation.

3. The third law is mentioned by Appian alone who says: "Now when the law of Gracchus had once been evaded by these tricks, an excellent law and most useful to the state if it could have been executed, another tribune not long after (οὐπολὺ ὕστερον) abolished even the vectigalia."[2] This is evidently the same law which Cicero mentions as that of Spurius Thorius and as he also mentions him in another place (*De Or.*, II, 70, 284), we may possibly accept him as the author.

There are still extant some fragments of a bronze tablet which contains upon its smooth surface the Lex Repetundarum and has cut upon its rough[3] back an agrarian law. These fragments were discovered in the 16th century among the collections in the Museum of Cardinal[4] Bembo at Padua. Sigonius attempted the reconstruction of this law and after him Haubold and Klentze, but Rudorff has completed the reconstruction as far as possible and made the law the subject of an interesting essay.[5] Mommsen has a commentary in the Corpus Inscriptionum Latinarum[6] upon this law. From all these sources the date of this law has been established almost beyond doubt as 111. Sigonius assigned it

---

[1] Corpus Inscriptionum Latinarum, vol. 1, p. 74.
[2] Appian, 1, c. 27.
[3] Long, I, 355; Wordsworth, 440.
[4] Long, I, 355; Wordsworth, 440; See Rudorff, Ack. des Sp. Thor.
[5] Zeitschrift für geschichtliche Rechtswissenschaft, Band X, s. 1-194.
[6] C. I. L., I, pp. 75-86.

to Spurius Thorius, and, as the name is immaterial and[1] his arguments moreover for this title are not easily set aside, we can do no better than adopt it.

Argument of the Lex Thoria.[2]

The law evidently consists of three parts, although the rubricae are absent.
 I. De agro publico p. R. in Italia (1–43).
 II. De agro publico p. R. in Africa (44–95).
 III. De agro publico p. R. qui Corinthorum fuit (96–105).

I. On the Ager Publicus in Italy.

This part may be divided roughly into three sections: (1) Lines 1–24, defining *ager privatus;* (2) 24–32, defining *ager publicus;* (3) 33–43, on disputed cases.

It thus embraces the first forty-three lines of the law, and is concerned with the public land of Italy, from the Rubicon southwards. It commences by referring to the condition of this land in the year 133, when Tiberius Gracchus was tribune. The law does not affect to touch any thing which had been enacted concerning this land prior to 133. It either confirms or alters what had been done in 133, and since that time. All the public land which was exempted from the operation of the Sempronian laws, *i. e.*, *Ager Campanus* and *Ager Stellatis*, was also excluded from the operation of the *lex Thoria*.

(1) The first ten lines of the law relate to that part of the ager publicus which was occupied before the time of the Gracchi, if the amount of such land did not exceed the maximum fixed by the Sempronian laws;

(2) Also, to the assignments made by lot (*sortito*) to Roman citizens by the commissioners since the enactment of the Sem-

---
[1] Long, I, 356.
[2] Wordsworth, 447. See the text of this law in C. I. L., vol. I, pp. 79–80.

pronian laws, if such assignments were not made out of land which had been guaranteed to the old possessors;

(3) Also, to all lands taken from an old possessor, but on his complaint restored to him by the commissioners;

(4) Also, to all houses and lands, in Rome or in other parts of Italy, which the commissioners had granted without lot, so as such grants did not interfere with the guaranteed title of older possessors;

(5) Also, to all the public land which Gaius Sempronius, or the commissioners, in carrying out his law, had used in the establishment of colonies or given to settlers, whether Roman citizens, Latini, or Italian Socii, or which they had caused to be entered on the "*formae*" or "*tabulae*."

All the lands comprised in the above are declared in lines seven and eight to be private property, in these words: "Ager locus omnis quei supra scriptus est, extra eum agrum locum, quei ager locus ex lege plebeivescito, quod C. Sempronius Ti. f. tr. pl. rogavit, exsceptum cavitumve est nei divideretur . . . . privatus esto."

Lines 8–10 declare that the censors shall, from time to time, enter this land upon their books like any other private property; and it is further declared that nothing shall be said or done in the senate to disturb the peaceful enjoyment of this land by those persons possessing it.

Of lines 11–13 (ch. II) nothing definite can be said, because of the few words which have been preserved.[1] Rudorff explains them as referring to land granted to *viasii vicani* (dwellers in villages along the roads), by the Sempronian commissioners; such lands to remain in their possession, but to be theoretically *ager publicus*.

Lines 13–14 refer to lands occupied since 133 *agri colendi causa*. They allow to every Roman citizen the privilege of occupying, for the purpose of cultivation, thirty jugera of public land; they further declare that he who shall possess or

---

[1] Long, I, 359.

have not more than thirty jugera of such land, shall possess and have it as private property,[1] with the provision that land so occupied shall be no part of the public land excepted from appropriation, and further, that such occupation shall not interfere with the guaranteed lands of a previous possessor.

Lines 14–15 relate to holders of pasture land (*ager compascuus*). This *ager compascuus* was land which had been left undivided, and had not become the private property of any individual, but was the common property of the owners of the adjacent lands. These persons had the right to pasture stock upon this land by paying pasture dues (*scriptura* or *vectigal*) to the state. The *Thoria lex* freed these lands from the *vectigal* or *scriptura*, and granted free pasturage to each man for ten head of large beasts—cattle, asses, and horses—and fifty head of smaller animals—sheep, goats, and swine. This common pasture must be carefully distinguished from the communal property which was granted to the settlers in a Colonia and called "*compascua publica*" with the additional title [2] of the colony, as "*Julienses*."

These rights of common resemble, in some respects, the English common of pasture as described by Bracton.[3] By English customary law, every freeholder holding land within a manor, had the right of common of pasturage on the lord's wastes as an incident to his land.

Lines 15–16. The possession of land, granted by the commissioners in a colony since 133, to be confirmed before the Ides of March next.

Lines 16–17. The same rule applied to lands granted otherwise by the same commissioners.

Line 18. Such occupants if forcibly ejected to be restored.

Lines 19–20. Land assigned by the Sempronian com-

---

[1] "Quom quis ceivis Romanus agri colendi causa in eum agrum agri jugera non amplius XXX possidebit habebitue, is ager privatus esto."
[2] Long, *loc. cit.*; Wordsworth, 446.
[3] Digby, *History of the Law of Real Property in England*, p. 157.

mission, in compensation for land in a colony which had been made public, to become private.

Lines 23–24. Confirmation of the title or restitution of such land to be made before the Ides of March next.

Lines 24–25. Land besides this which remains public is not to be occupied, but to be left free to the public for grazing. A fine for occupation is imposed. The law allowed all persons to feed their beasts great and small on this public pasture, up to the number mentioned in lines 14–15 as the limit to be pastured on the *ager campascuus*, free of all tax. This, according to Rudorff, was done for the benefit of the small holders. Those who sent more than this number of animals to the public pastures must pay a *scriptura*, for each head.

Line 26. While the cattle or sheep were driven along the '*calles*,' or beast-tracks, and along the public roads to the pasture grounds, no charge was made for what they consumed along the road.

Line 27. Land given in compensation out of public land, to be *privatus utei quoi optuma lege*.

Line 27. Land taken in this way from private ownership to be *publicus*, as in 133.

Lines 27–28. Land given in compensation for *ager patritus* to be itself *patritus*.

Line 28. Public roads to remain as before.

Line 29. Whatever Latins and *peregrini* might do in 112, and whatever is not forbidden citizens to do by this law, they may do henceforward.

Lines 29–30. Trial of a Latin to be the same as for a Roman citizen.

Lines 31–32. Territory (1) of borough towns or colonies (2), in trientabulis, to be, as before, public.

Lines 33–34. Cases of dispute about land made private between 133 and 111, or by this law, to be judged by the consul or practor before next Ides of March.

Lines 35–36. Cases of dispute after this date to be tried by consuls, practors, or censors.

Lines 36-39. Judgment on money owing to publicani to be given by consuls, proconsuls, praetors or propraetors.

Line 40. No one to be prejudiced by refusing to swear to laws contrary to this law.

Lines 41-42. No one to be prejudiced by refusing to obey laws contrary to this law.

Lines 43-44. On the colony of Sipontum (?).

Thus we see that the *lex Thoria* had two main objects in view: (1) The guaranteeing to possessors full property in the land which they occupied. (2) The freeing from *rectigal* or *scriptura* the property of every one.

In this way was the reaction of the aristocracy completed. It left nothing of the Sempronian law. Appian[1] has fully comprehended all this, and, in his enumeration of the three laws, connection between which he indicates, we see clearly the entire revolutionary system, conducted, we must admit, with a rare address and a perfidy which rendered the effect certain. The aristocracy did not rest. As soon as they had gained the people by their new bait of money and food, soothed them by their apparent generosity, and familiarized them with the idea that the *possessions* of the nobles were not only legally acquired but inviolable, then they raised the mask, and by a bold step swept away the *rectigal*,[2] thus leaving their property free. The enactment of this law virtually closed the long struggle between patrician and plebeian over the public lands of Rome, and left them as full property in the hands of the rich nobility. The results could hardly have been otherwise. Sumptuary laws, false economic principles, had closed all channels[3] of trade and manufacture to the nobility, while conquest had filled their hands with gold and placed at their disposal vast numbers[4] of slaves. There was but one channel open for the investment of this gold,—the agrarian.[5] Farming and cattle-raising were the only occupations in which

---

[1] Long, I, 357.   [2] Appian, I, c. 27.   [3] Long, *loc. cit.*; Ihne, *loc. cit.*
[4] Ihne, *loc. cit.*; Long, *loc. cit.*   [5] Momm., *loc. cit.*

slaves could be used with advantage and so, as a natural result of Roman economics, the plebeian, with little or no money and subject to the military call, was compelled to enter into a one-sided contest with capital and slave labor. So long as these conditions existed so long would all the laws of the world fail to save him from abject poverty and its attendant evils.

SEC. 14.—AGRARIAN MOVEMENTS BETWEEN 111 AND 86.

In the year following the enactment of the *lex Thoria*, or, by some other authorities, in 105, an agrarian law was proposed by a tribune named Marcus Philippus. Cicero is the only writer who mentions it, and he has given us no information concerning its tendency and dispositions. We only know from him that it was rejected.[1] Probably the whole thing was merely a political ruse in order to gain an election or to be handsomely bought off by the nobility. It, however, presents one point of interest to us. The introduction of the bill was preceded by a speech, in which the tribune, in justifying his undertaking, affirmed that there were not two thousand citizens who had wealth. Cicero has made no attempt to refute this, and must, therefore, have judged it true. It reveals the fact that Rome was in a deplorable condition.

In chronological order the first agrarian law after the vain attempt of Philippus was that of Lucius Appuleius Saturninus. In the year 100, he brought forward a bill for the distribution of land in Africa[2] to the soldiers of Marius. Each soldier was to receive one hundred jugera of land. No distinction was to be made between Roman and Latin. This

---

[1] Cic., *De Off.*, II, 21.

[2] Lucius Appuleius Saturninus, tribunus plebis seditiosus ut gratiam Marianorum militum pararet, legem tulit ut veteranis centena agri jugera in Africa dividerentur .... Siciliam, Achaiam, Macedoniam novis colonis destinavit; et aurum, dolo an scelere, Caepionis partum, ad emtionem agrorum convertit. Aurel. Victor. De Vir. Illus., 73.

bill received the sanction of the assembly and became a law, but force was the chief instrumentality in bringing this about. This law, so far as can be ascertained, was never enforced, so that when the same man, three years later, brought forward another agrarian bill, he took the precaution to add a clause binding every senator, under heavy penalty, to confirm the law by the most solemn oath.[1] The first law was enacted in order to provide the soldiers of Marius with suitable farms when they returned from the campaign in Numidia. The author doubtless acted with the aid and hearty coöperation of Marius. When Saturninus brought forward his second bill, Marius[2] had returned from the north as the hero of Aquae Sextiae and was present to help. The nobility as one man opposed the scheme; the town-people were the clients of the rich. If Marius[3] and Saturninus were to succeed, it must be by the aid of the country burgess and the soldier. With the legions that fought at Vercellae drawn up in the town, amid riot and bloodshed, the assembly passed the bill. The senate, together with Marius himself, for a time demurred from taking the oath. Finally,[4] at the instigation of "the man from the ranks," who had come to the conclusion that it was best to subscribe, all save one, Metellus, took the oath. The law enacted that assignments of land in the country of the Gauls, in Sicily, Achaia, and Macedonia, should be made; that colonies should be established, and that Marius should be the head of the commission entrusted with the establishment of all these settlements.[5] These colonies were to consist of Roman citizens; and, in order that Latini,[6] their companions in arms, might participate in the grants, Marius was invested with power to bestow the franchise upon a certain number of these. But no one of these colonies was ever founded. The only

---

[1] App., I, 29; Plutarch, *Marius*, 29.
[2] Plutarch, *Marius*, *loc. cit.*
[3] App., *Bell. Civ.*, I, 30–33.
[4] App., *loc. cit.*
[5] Aurelius Victor, 73.
[6] Cicero, *De Orat.*, II, c. 7, 1; *pro Balbo*, XIV; *pro Rabirio*, XI.

colony of the year 100 was Eporedia[1] (Ivrea), in the northwestern Alps, and it is not likely that this was established in accordance with the provisions of the enactment. The law was to take effect in 99, and a change of party took place before that time which sent Marius into practical banishment and rewarded his partisan, Saturninus, with death. The optimates who were now in office paid no attention to the law, and the senators forgot their oath. Another injury is added to the many which the Latini had suffered.

In the year 99, *i. e.*, in the year following the death of Saturninus, an agrarian law was proposed by the tribune Titius, but we know nothing of its conditions. Cicero is the only writer who mentions it and even his text is doubtful.[2] According to one of his statements Titius was banished because he had preserved a portrait of Saturninus, and the knights deemed him for this reason a seditious citizen. Valerius Maximus, who without doubt borrowed his facts from Cicero, states that "Titius had rendered himself dear to the people by having[3] brought forward an agrarian law." Cicero mentions in another place, the *lex Titia*[4] upon the same page as the *lex Saturnina* and implies that it had been enacted. If so it was disregarded and thus rendered void.

In 91 an agrarian law was proposed by Livius Drusus, the son of the adversary of Gaius Gracchus, and, with his new judiciary, the measure was carried and became a law.[5] The Italians were embraced in this law and were to have equal rights with Roman citizens, but Drusus died before he had time to carry his law into execution, and his law died with him.

---

[1] Long, I.   [2] Cicero, *Pro Rabirio*, 9.
[3] Val. Max., VIII, 1, § 2: "Sext. Titius . . . agraria lege lata gratiosus apud populum."
[4] *De Legibus*, II, 6. *De Orat.*, II, 11.
[5] Ihne, V, 176-186; App., I, 35; Val. Max., IX, 5, 2: Cicero, *De Orat.*, III, 1; Livy, *Epit.*, 71.

SEC. 15.—EFFECT OF THE SULLAN REVOLUTION.

As soon as Sulla found himself established, he caused a bill to pass the Comitia Centuriata by means of which he was empowered to inflict punishment upon certain Italian communities. For the accomplishment of this purpose commissioners were appointed to coöperate with the garrisons established throughout all Italy. The less guilty were required to pay fines, pull down their walls, and raze their citadels.[1] Those that had been guilty of continued opposition, as Samnium, Lucania, and Etruria, had their territory in whole or in part confiscated, their municipal rights cancelled, immunities taken from them, which had been granted by old treaties, and the Roman franchise,[2] which they had been granted by the Cinnan government, annulled. Such persons received, instead, the lowest Latin rights which did not even imply membership in any community and rendered them destitute of civic constitution and the right of making a testament.[3] This latter treatment applied only to those whose land was confiscated. Thus Sulla vindicated the majesty of the Republic and at the time avoided furnishing his enemies with a nucleus in Italian communities. In Campania, the democratic colony established at Capua by Cinna[4] was done away with and the domain given back to the state, thus becoming *ager publicus*. The whole territory of Praeneste and Norba in Latium, and Spoletium in Umbria was confiscated. The town of Sulmo in Pelignium was razed. But more direful than all this was the punishment which fell upon Etruria[5] and Samnium. These people had marched upon Rome and, with the avowed determination of

---

[1] App., *Bell. Civ.*, 1, 94–100; Livy, *Epit.*, 89. Plutarch, *Life of Sulla*.
[2] Ihne, V, 391.
[3] Momm., III, 428, note. See article on Sulla, in Brittannica.
[4] Momm., III, 401.
[5] Momm., III, 429; Ihne, V, 392; Long.

exterminating the Roman people, had engaged in battle at the Colline gate. They were utterly destroyed and their country left desolate. The territory of Samnium was not even opened up for settlement, but left as a lair for wild beasts. Henceforth from the Rubicon to the Straits of Sicily there were to be none but Romans; the laws and the language of the whole peninsula were to be the laws[1] and the language of Rome.

To accomplish such an object as this, it was not enough to destroy and make desolate, it became necessary to repopulate the waste places and rebuild that which had been torn down. Roman citizens had to be sent as colonists into the desolate regions. Sulla, accordingly, undertook to carry out his plans of colonization, the grandest and most comprehensive which Rome had ever seen, and which indeed have had no parallel in history till the settlement of the north of Ireland by Cromwell and William III. The arrangements as to the property of the Italian soil placed at the disposal of Sulla[2] all the Roman domain lands which had been placed in usufruct to the allied communities, and which now reverted to the Roman government. It also placed at his disposal all the confiscated territories of the communities incurring punishment. Upon these territories he established military colonies, and thus obtained a three-fold result.[3] He remunerated his soldiers for the faithful service rendered him in long years of toil and danger. He repeopled the regions desolated by war (except Samnium). He provided a military protection for himself and the new constitution which he established.

Most of his new settlements were directed to Etruria, Faesulae and Arretium being among the number; others, to Latium[4] and Campania, where Praeneste and Pompeii became Sullan

---

[1] Momm., III, 429. [2] Momm., *loc. cit.*; Ihne, V, 391-395.
[3] Momm., III, 429.
[4] Momm., III, 430; Marquardt u. Momm., *Röm. Alter.*, IV, 111, totam Italiam suis praesidiis obsidere atque ocupare; Cicero, *De Leg. Agr.*, 2, 28, 75.

colonies. A great part of these colonies were, after the Gracchan manner, merely grafted upon town-communities already existing. The comprehensiveness of these settlements may be seen in this fact that 20,000 allotments were [1] made in different parts of Italy. Notwithstanding this vast disposal of territory, Sulla gave lands to the temple of Diana at Mt. Tifata, while the territory of Volaterrae and Arretium remained undisturbed. He also revived the old plan of occupation which had been legally forbidden in the year 118. Many of Sulla's intimate friends availed themselves of this method of becoming masters of large estates.

SEC. 16.—AGRARIAN MOVEMENTS BETWEEN 86 AND 59.

The first agrarian movement after the Sullan Revolution was that inaugurated by the tribune Rullus. This has become the most famous of all the agrarian laws because of the speeches made against it by the great adversary of Rullus, Cicero, who succeeded in defeating the measure by reason of his brilliant rhetoric. Plutarch[2] has thus analyzed this proposition. "The tribunes of the people proposed dangerous innovations; they demanded the establishment of ten magistrates with absolute power, who, while disposing, as masters, of Italy, Syria, and the new conquests of Pompey, should have the right to sell the public lands; to prosecute those whom they wished; to banish; to establish colonies; to draw upon the public treasury for whatever money they had need; to levy and maintain what troops they deemed necessary. The concession of so widely extended power gained for the support of the law the most powerful men in Rome. The colleague of Cicero, Antonius, was one of the

---

[1] App., I, 100; Cicero, *De Legibus Agrariis*, II, 28, 78; Ihne, V, 394; Marquardt u. Momm., IV, 111; Zumpt, *Comm. Epigr.*, 242–246; Cicero, *Ad Att.*, I, 19, 4: "Volaterranos et Arretinos, quorum agrum Sulla publicarat."
[2] Plutarch, *Cicero*, 16–17.

first to favor it, in the hope of being one of the decemvirs. Cicero opposed the new law in the senate and his eloquence so completely overpowered even the tribunes that they had not one word to reply. But they returned to the charge and having gained the support of the people, they brought the matter before the tribes. Cicero was in no way alarmed; he left the senate, appeared on the rostrum before the people and spoke with so great force that he not only caused the law to be rejected but took from the tribunes all hope of being successful in similar enterprises."

In 61 we find Cicero advocating a bill similar in nature to the one he had so brilliantly combatted in 64. In the last instance, however, the law was proposed by Pompey, and in favor of Pompey's soldiers and that made all difference to a man who ever curried favor with the great. Flavius, who proposed this law, was but the creature of Pompey. Cicero has made known to us, in one of his letters to Atticus, the conditions of the law which Flavius proposed and the modifications which he himself wished to apply to it. Flavius proposed to distribute lands both to the soldiers of Pompey and the people; to establish colonies; to use for the purchase of the lands for colonization, the subsidies which should accrue in five years, from the recently conquered territories.[1] The senate rejected this law entirely, in the same spirit of opposition which it had shown to all agrarian laws, probably thinking that Pompey would thereby obtain too great an increase of power.[2] This was the last attempt at agrarian legislation until the year 59, when Julius Cæsar enacted his famous law.

---

[1] Cicero, *Ad. Att.*, I, 19.
[2] Ibid.: "Huic toti rationi agrariae senatus adversabatur, suspicans Pompeio novam quamdam potentiam quaeri."

Sec. 17.—Lex Julia Agraria.

During the first consulship of Caius Julius Cæsar, he brought forward an agrarian[1] bill at the instigation of his confederates. The main object of this bill was to furnish land to the Asiatic army[2] of Pompey. In fine, this bill was little more than a renewal of a bill presented by Pompey the previous year (58), but rejected. Appian gives the following account of this bill: "As soon as Cæsar and Bibulus[3] (his colleague) entered on the consulship, they began to quarrel and to make preparation to support their parties by force. But Cæsar who possessed great powers of dissimulation, addressed Bibulus in the senate and urged him to unanimity on the ground that their disputes would damage the public interests. Having in this way obtained credit for peaceable intentions, he threw Bibulus off his guard, who had no suspicion of what was going on, while Cæsar, meanwhile, was marshalling a strong force, and introducing into the senate laws for favoring the poor, under which he proposed to distribute land among them and the best land in Italy, that about[4] Capua which at the present time was let on public account.[5] He proposed to distribute this land among heads of families who had three children, by which measure he could gain the good will of a large multitude, for the number of those who had three children was 20,000. This proposal met with opposition from many of the senators, and Cæsar, pretending to be much vexed at their unfair behavior, left the house and never called the senate together again during the

---

[1] Livy, *Epit.*, 103.   [2] Momm., IV, 244.
[3] App., *Bell. Civ.*, II, c. 10.
[4] Compare Dio Cassius, Bk., XXXVIII, c. 1: "Τὴν δὲ χώραν τὴν δὲ κοινὴν ἅπασαν πλὴν τῆς Καμπανίδος ἔνεμε, ταύτην γὰρ ἐν τῷ δημοσίῳ ἐξαίρετον διὰ τὴν ἀρέτην συνεβούλευσεν εἶναι."
[5] Compare Suetonius' *Cæsar*, c. 20: "Campum Stellatem, majoribus consecratum, agrumque Campanum, ad subsidea reipublicae vectigalem relictum."

remainder of his consulship, but addressed the people from the rostra. He, in the presence of the assembly, asked the opinion of Pompeius and Crassus, both of them approving, and the people came to vote on them (the bills), with concealed daggers. Now as the senate[1] was not convened, for one consul could not summon the senate without the consent of the other consul, the senators used to meet at the house of Bibulus, but they could make no real opposition to Cæsar's power. . . . Now Cæsar secured the enactment of the laws, and bound the people by an oath to the perpetual observance of them, and he required the same oath from the senate. As many of the senators opposed him, and among them Cato, Cæsar proposed death as a penalty for not taking the oath and the assembly ratified this proposal. Upon this all took the oath immediately because of fear, and the tribunes also took it, for there was no longer any use in making opposition after the proposal was ratified."

This agrarian law did not affect the existing rights of property and heritable possession. It destined for distribution only the Italian domain land, that is to say, merely the territory of Capua, as this was all that belonged to the state.[2] If this was not enough to satisfy the demand, other Italian lands were to be bought out of the revenue from the eastern provinces at the taxable value rated in the censorial rolls. The number of persons settled on the *Campanus ager* is said[3] to have been 20,000 citizens who had each three children or more. The land was not distributed by lot, but at the pleasure of the commissioners, each one receiving some 30 jugera.[4] If 20,000 heads of families with their wives and three children in each family were settled in Campania, the whole number of settlers would be 100,000. This great number could scarcely leave Rome at one time, and we find

---

[1] App., II, c. 11.
[2] App., II, c. 20, and Suetonius, *Julius Cæsar*, c. 20.
[3] Suetonius, *loc. cit.*   [4] Lange, *Röm. Alter.*, III, 273.

that as late as 51 the land was not all assigned.[1] While the tenor of the law does not imply that it was the intention to reward military service with grants of land, yet we may be sure that the veterans of Pompey were not forgotten.[2] There are no extant authorities which speak of the settlement of the Campanian land that say any thing about the soldiers settled there, unless it be Cicero. He speaks of the Campanian territory being taken out of the class that contributed a revenue to the state in order that it might be given to soldiers,[3] and he appears to refer to this time (59). Mommsen says that "the old soldiers as well as the temporary lessees to be ejected were simply recommended to the special consideration of the land distributors."[4] These latter were a commission of twenty appointed by the state. Cæsar, at his own request, was excused from serving, but Pompey and Crassus were the chief ones, thus furnishing sufficient reason for supposing that the soldier was provided for. The passage of this bill amounted in substance to the reëstablishment of the democratic colony founded by Marius and Cinna and afterwards abolished by Sulla.[5] Capua now became a Roman colony after having had no municipal constitution for one hundred and fifty-two years, when the city with all its dependencies was made a prefecture administered by a prefect of Rome. The revenues from this district were doubtless no longer needed, as those from Pontus and Syria[6] supplied all the needs of the government, but it is difficult to see what benefit could be reaped from the ejection of the thrifty farmers who, as tenants of the state, cultivated this territory and paid their rents regularly into the state coffers.

---

[1] Cicero, ad Att., VIII, 4.
[2] Dion Cassius, 45, c. 12; Cicero, ad Att., X, 8.
[3] Cicero, Phil., II, 39: "agrum Campanum, qui cum de vectigalibus eximebatur, ut militibus daretur." Marquardt u. Momm., Röm. Alter. IV, 114.
[4] Momm., IV, 244.
[5] Momm., III, 392, 428.
[6] Momm., III, 392, 428.

Wherever the new settlers were brought in, the old cultivators were turned out. No ancient writer says anything about the condition of these people. Cicero, in his second speech upon the land bill of Rullus, when speaking of the consequences that would follow its enactment, declared that if the Campanian cultivators were ejected they would have no place to go, and he truly says that such a measure would not be a settlement of plebeians upon the land, but an ejection and expulsion of them from it.[1]

Did it pay to send out a swarm of 100,000 idle paupers[2] who, for two generations, had been fed at the public charge from the corn-bins of Rome, simply in order that a like number of honest peasants, who had been not only self-supporting but had paid a large part of the Roman revenue, should be compelled to sacrifice their goods in a glutted market and become debauched and idle?

Sec. 18.—Distribution of Land after the Civil War between Cæsar and Pompey.

After Pompey had been vanquished at Pharsalia, and the republicans in Africa, Cæsar proceeded to distribute lands to his soldiers in accordance with his promise to give them lands, "not by taking them from their proprietors as Sulla did; not by mixing colonists with citizens despoiled of their goods and thus breeding perpetual strife,—but by dividing both public land and his own private property,[3] and, if this were not sufficient, by buying what was needed." Appian says that Cæsar did not succeed in carrying out these promises in full, but that veterans were in some cases settled upon lands legally belonging to others.[4] However, his soldiers were not huddled together like those of Sulla, in military colonies of

---

[1] Cicero, *Rul.*, II, c. 31.   [2] Cicero, *Phil.*, II, 17.
[3] App., 94.   [4] App., II, 120.

their own, but when they settled in Italy they were scattered[1] as much as possible throughout the entire peninsula in order to make them more easily amenable to the laws.[2] In Campania, where Cæsar had lands at his disposal, the soldiers were settled in colonies, and so, close together. According to a letter of Cicero to Paetus, among the lands distributed were those of Veii and Capena. Historians have estimated that there were 100,000 soldiers who received lands in Italy by this distribution.

SEC. 19.—DISTRIBUTIONS FROM THE DEATH OF CÆSAR TO THE TIME OF AUGUSTUS.

The death of Cæsar in no way stopped the assignment of lands, but rather rendered all possession of land in Italy unsafe. A few weeks after his death two new laws were promulgated, one by the tribune, Lucius Antonius,[3] a *lex agraria*, and the other the *lex de colonis in agros deducendis* by the consul Marcus Antonius. The first was enacted on the 5th of June,[4] and ordered that all the *ager publicus* still at the disposal of the state, including the Pomptine marshes which Cæsar had at one time planned to drain, but had not, be divided among the veterans and citizens. It was abrogated by a *senatus consultum* of the 4th of January, 43,[5] but was nevertheless carried into execution almost immediately with great relentlessness towards the enemies[6] of Antonius. The second, the *Lex Antonia*, perished in April of 44, and had as a result the establishment of a colony near Casilinum,[7] which Cæsar had already colonized; the remainder of the domain

---

[1] Long; Momm.   [2] Suetonius, *Julius Cæsar*, 38.
[3] L. Langii, Commentationis de Legibus Antoniis a Cicerone Phil., V, 4, 10; Commemoratis particula prior et posterior; Lipsiae, 1882; Lange, *Röm. Alter.*, III, 499, 503, 526; Marquardt u. Momm., *Röm. Alter.*, IV, 116.
[4] Lange, *Comm.*, II, 14.   [5] Cicero, *Phil.*, VI, 5, 14; XI, 6, 13.
[6] *Phil.*, V, 7, 20.   [7] Langii, *Comm.*, II, 14.

lands, the *ager Campanus* and *ager Leontinus*, was converted into a reward for the supporters of Antonius.[1] This was also set aside by the new law of the consul C. Vibius Pansa, in February, 43.[2]

Second Triumvirate. When Antony, Lepidus, and Octavius were reconciled, thus forming the second triumvirate, the treaty sanctioning this new state of affairs stipulated, in favor of the soldiers, a new distribution of lands, *i. e.*, a new agrarian law; Appian says:—"In order to increase the zeal of the army, the triumvirs promised to the soldiers, independent[3] of other results of victory and a gratuity of colonies, 18 Italian towns, important by means of their wealth and the richness of their lands. These were divided among the soldiers with their lands and buildings, as conquered towns. Among the number were Capua, Rhegium, Venusia, Beneventum, Nuceria and Vibo. Thus the most beautiful part of Italy became the prey of the soldiers."

Dion Cassius, Suetonius and Velleius Paterculus all mention these assignments. After the battle of Philippi and the defeat and death of Brutus and Cassius, 170,000 men were provided for, in accordance with these promises, out of the goods of the proscribed and the lands confiscated to the state. The lands of the towns mentioned in Appian were taken under the form of a forced sale, but the purchase money was never paid owing to the bankrupt condition of the treasury.

If we examine into the nature of these agrarian laws since the death of Julius Cæsar, we shall find that they differ in all respects from previous enactments:

1. They were executed at the expense not only of public domains but also of private property.

---

[1] Cic., *Phil.*, II, 17, 43; II, 39, 101; III, 9, 22; VIII, 8, 26; Dio Cass., 45, 30; 46, 8.

[2] Cic., *Phil.*, V, 4, 10; V, 19, 53; X, 8, 17; VIII, 15, 31.

[3] "Δόσεσι τῶν Ἰταλικῶν πόλεων ὀκτωκαίδεκα . . . ὥσπερ αὐτοῖς ἀντὶ τῆς πολεμίας δορίληπτοι γενόμεναι. . . . Οὕτω μὲν τὰ κάλλιστα τῆς Ἰταλίας τῷ στρατῷ διέγρεφον." App., IV, 3.

2. They were the work of one man and not of the entire people.

3. The name of the people was never mentioned in these laws; they were enacted wholly for the profit of the soldiery. Before the distributions made by the triumvirate, the public lands had been absorbed, or at least the fragments remaining were in no way sufficient to recompense the service of the veterans.

Upon the establishment of the empire, the public lands became a vast manorial estate whose over-lord was the emperor himself.

FINIS.

THE JOHNS HOPKINS PRESS
BALTIMORE

I. **American Journal of Mathematics.** S. NEWCOMB, Editor, and T. CRAIG, Associate Editor. Quarterly. 4to. Volume XIII in progress. $5 per volume.

II. **American Chemical Journal.** I. REMSEN, Editor. 8 nos. yearly. 8vo. Volume XIII in progress. $4 per volume.

III. **American Journal of Philology.** B. L. GILDERSLEEVE, Editor. Quarterly. 8vo. Volume XII in progress. $3 per volume.

IV. **Studies from the Biological Laboratory.** H. N. MARTIN, Editor, and W. K. BROOKS, Associate Editor. 8vo. Volume V in progress. $5 per volume.

V. **Studies in Historical and Political Science.** H. B. ADAMS, Editor. Monthly. 8vo. Vol. IX in progress. $3 per volume.

VI. **Johns Hopkins University Circulars.** 4to. Vol. X in progress. $1 per year.

VII. **Johns Hopkins Hospital Bulletin.** 4to. Monthly. $1 per year.

VIII. **Johns Hopkins Hospital Reports.** 4to. $5 per year.

IX. **Contributions to Assyriology, etc.** Vol. I ready. $8.

X. **Annual Report of the Johns Hopkins University.** Presented by the President to the Board of Trustees.

XI. **Annual Register of the Johns Hopkins University.** Giving the list of officers and students, and stating the regulations, etc. *Published at the close of the academic year.*

ROWLAND'S PHOTOGRAPH OF THE NORMAL SOLAR SPECTRUM. New edition now ready. Set of ten plates, mounted. $20.

THE OYSTER. By William K. Brooks. 240 pp., 12mo. 14 plates. $1.00.

THE TEACHING OF THE APOSTLES (complete facsimile edition). J. Rendel Harris, Editor. 110 pp. and 10 plates. 4to. $5.00, cloth.

OBSERVATIONS ON THE EMBRYOLOGY OF INSECTS AND ARACHNIDS. By Adam T. Bruce. 46 pp. and 7 plates. $3.00, cloth.

SELECTED MORPHOLOGICAL MONOGRAPHS. W. K. Brooks, Editor. Vol. I. 370 pp. and 51 plates. 4to. $7.50, cloth.

REPRODUCTION IN PHOTOTYPE OF A SYRIAC MS. WITH THE ANTILEGOMENA EPISTLES. I. H. Hall, Editor. $3, paper; $4, cloth.

STUDIES IN LOGIC. By members of the Johns Hopkins University. C. S. Peirce, Editor. 123 pp., 12mo. $2.00, cloth.

NEW TESTAMENT AUTOGRAPHS. By J. Rendel Harris. 54 pp. 8vo; 4 plates. 50 cents.

THE CONSTITUTION OF JAPAN, with Speeches, etc., illustrating its significance. 48 pp., 16mo. 50 cents.

ESSAYS AND STUDIES. By Basil L. Gildersleeve. 520 pp., small 4to. $3.50, cloth.

A full list of publications will be sent on application.

Communications in respect to exchanges and remittances may be sent to The Johns Hopkins Press, Baltimore, Maryland.

JOHN MURPHY & CO'S NEW PUBLICATIONS.

A BOOK FOR THE TIMES.

OUR CHRISTIAN HERITAGE.

By His Eminence Cardinal Gibbons.

One Volume.   *12mo.*   *Cloth.*   *524 pages.*   *Price $1.00.*

It is a noble work, in execution as well as conception.—*New York Herald.*
The Agnostic, as he lays down the book, will be inclined to say, with King Agrippa, "Almost thou persuadest me to be a Christian."—*New York Sun.*

SECOND REVISED EDITION.

HEREDITY.

By W. K. BROOKS, Johns Hopkins University.

One Volume.   *12mo.*   *Cloth.*   *Price $2.00.*

This work combines in a very unusual degree the two traits that are so rarely found to coexist in scientific books: it is both original and independent in its views, and is at the same time a most lucid and popular presentation of its subject.—*Popular Science Monthly.*

1791; A TALE OF SAN DOMINGO.

An Authentic Account of the Uprising of the Negroes in 1791 which destroyed the famous old French Colony.

By E. W. GILLIAM, M. D.

1 Volume.   *12mo.*   *Cloth.*   *Net, $1.*

The study of the situation and a quiet but effective style makes "1791" a story of more than ordinary interest.—*Baltimore Sun.*
The material for this most interesting story was gathered from facts connected with the slave insurrection on the island of San Domingo in 1791.—*Telegraph.*

SCIENCE AND SCIENTISTS.

Some Papers on Natural History, by the Rev. JOHN GERARD, S. J.

Contents:—Mr. Grant Allen's Botanical Fables—Who Painted the Flowers—Some Wayside Problems—Behold the Birds of the Air—How Theories are Manufactured—Instinct and its Lessons.

1 Volume.   *12mo.*   *Cloth, 40 cents net.*

The author starts out by refuting some of Grant Allen's botanical assertions, and he does it so neatly that the most interesting English naturalist is literally knocked down with a feather. He also, very frankly, attacks the evolutionists and Darwinians, but giving us no more plausible theories as to who painted the flowers and the berries, or who taught the birds to sing, except that, "These are Thy glorious works, Parent of good."

JOHN MURPHY & CO., Publishers, Baltimore.

PERIODICALS PUBLISHED BY

FELIX ALCAN

108 Boulevard Saint-Germain,

PARIS.

Revue Historique.
Edited by M. G. MONOD, Lecturer at the Ecole Normale Supérieure, Adjunct Director of the Ecole des Hautes Etudes.

16th Year, 1891.

The "Revue Historique" appears bi-monthly, making at the end of the year three volumes of 500 pages each.
Each number contains: I. Several leading articles, including, if possible, a complete thesis. II. Miscellanies, composed of unpublished documents, short notices on curious historical points. III. Historical reports, furnishing information, as complete as possible, touching the progress of historical studies. IV. An analysis of periodicals of France and foreign countries, from the standpoint of historical studies. V. Critical reports of new historical works.
By original memoirs in each number, signed with the names of authorities in the science, and by reports, accounts, chronicles and analysis of periodicals, this review furnishes information regarding the historical movement as complete as is to be found in any similar review.
Earlier series are sold separately for 30 frs., single number for 6 frs., numbers of the first year are sold for 9 frs.
Price of subscription, in Postal Union, 33 frs.

Annales de l'Ecole Libre des Science Politiques.
Published tri-monthly by the Coöperation of Professors and Former Pupils of the College.

6th Year, 1891.

Committee of publication: MM. BOUTMY, Director of the College; LÉON SAY, Member of the Académie Française, formerly Minister of Finance; A. DE FOVILLE, Professor at the Conservatory of Arts and Trades, Chief of the Bureau of Statistics in the Ministry of Finance, (Treasury Department); R. STOURM, formerly Inspector of the Finances and Administrator of Indirect Taxes; AUG. ARNAUNÉ; A. RIBOT, Deputy; GABRIEL ALIX; L. RENAULT, Professor at the Law College of Paris; ANDRÉ LEBON, Chief of the Cabinet of the President of the Senate; ALBERT SOREL; PIGEONNEAU, Substitute Professor at the College de Paris; A. VANDAL, Auditor of the First Class.
The subjects treated include the whole field covered by the programme of instruction: Political Economy, Finance, Statistics, Constitutional History, Public and Private International Law, Law of Administration, Comparative Civil and Commercial Legislation, Legislative and Parliamentary History, Diplomatic History, Economic Geography, Ethnography. The Annals besides contain Bibliographical Notices and Foreign Correspondence.
Subscription in Postal Union, 19 frs.

Revue Phiosophique de la France et de l'Etranger.
Edited by TH. RIBOT, Professor at the College of France.

16th Year, 1891.

The "Revue Philosophique" appears monthly, and makes at the end of each year two volumes of about 680 pages each.
Each number of the "Revue" contains: 1. Essays. 2. Accounts of new philosophical publications, French and foreign. 3. Complete accounts of periodicals of foreign countries as far as they concern philosophy. 4. Notes, documents, observations. The earlier series are sold separately at 30 frs. and at 3 frs. by number. In Postal Union, 33 frs. Subscriptions to be paid in advance.

Payment may be for the periodicals through postal orders. The publisher will allow all expenses for money orders to be charged to him.

THE AMERICAN JOURNAL OF ARCHÆOLOGY
AND OF THE
HISTORY OF THE FINE ARTS.

The Journal is the organ of the Archæological Institute of America, and of the American School of Classical Studies at Athens, and it will aim to further the interests for which the Institute and the School were founded. It treats of all branches of Archæology and Art History: Oriental, Classical, Early Christian, Mediæval and American. It is intended to supply a record of the important work done in the field of Archæology, under the following categories: I. Original Articles; II. Correspondence from European Archæologists; III. Reviews of Books; IV. Archæological News, presenting a careful and ample record of discoveries and investigations in all parts of the world; V. Summaries of the contents of the principal archæological periodicals.

The Journal is published quarterly, and forms a yearly volume of about 500 pages royal 8vo, with colored, heliotype and other plates, and numerous figures, at the subscription price of $5.00. Six volumes have been published.

It has been the aim of the editors that the Journal, beside giving a survey of the whole field of Archæology, should be international in character, by affording to the leading archæologists of all countries a common medium for the publication of the results of their labors. This object has been in great part attained, as is shown by the list of eminent foreign and American contributors to the three volumes already issued, and by the character of articles and correspondence published. Not only have important contributions to the advance of the science been made in the original articles, but the present condition of research has been brought before our readers in the departments of Correspondence, and Reviews of the more important recent books. Two departments in which the Journal stands quite alone are (1) the *Record of Discoveries*, and (2) the *Summaries of Periodicals*. In the former a detailed account is given of all discoveries and excavations in every portion of the civilized world, from India to America, especial attention being given to Greece and Italy. In order to insure thoroughness in this work, more than sixty periodical publications are consulted, and material is secured from special correspondents.

In order that readers should know everything of importance that appears in periodical literature, a considerable space has been given to careful summaries of the papers contained in the principal periodicals that treat of Archæology and the Fine Arts. By these various methods, all important work done is concentrated and made accessible in a convenient but scholarly form, equally suited to the specialist and to the general reader.

All literary communications should be addressed to the managing editor,

A. L. FROTHINGHAM, Jr.,
PRINCETON, N. J.

All business communications to the publishers,

GINN & COMPANY.,
BOSTON, MASS.

MODERN LANGUAGE NOTES

A MONTHLY PUBLICATION
With intermission from July to October inclusive.

DEVOTED TO THE INTERESTS
OF THE
ACADEMIC STUDY OF ENGLISH, GERMAN,
AND THE
ROMANCE LANGUAGES.

A. MARSHALL ELLIOTT, *Managing Editor.*

JAMES W. BRIGHT, H. C. G. VON JAGEMANN, HENRY ALFRED TODD,
Associate Editors.

This is a successful and widely-known periodical, managed by a corps of professors and instructors in the Johns Hopkins University, with the co-operation of many of the leading college professors, in the department of modern languages, throughout the country. While undertaking to maintain a high critical and scientific standard, the new journal will endeavor to engage the interest and meet the wants of the entire class of serious and progressive modern-language teachers, of whatever grade. Since its establishment in January, 1886, the journal has been repeatedly enlarged, and has met with constantly increasing encouragement and success. The wide range of its articles, original, critical, literary and pedagogical, by a number of the foremost American (and European) scholars, has well represented and recorded the recent progress of modern language studies, both at home and abroad.

The list of contributors to MODERN LANGUAGE NOTES, in addition to the Editors, includes the following names:—

ANDERSON, MELVILLE B., State University of Iowa; BANCROFT, T. WHITING, Brown University, R. I.; BASKERVILL, W. M., Vanderbilt University, Tenn.; BOCHER, FERDINAND, Harvard University, Mass.; BRADLEY, C. B., University of California, Cal.; BRANDT, H. C. G., Hamilton College, N. Y.; BROWNE, WM. HAND, Johns Hopkins University, Md.; BURNHAM, WM. H., Johns Hopkins University, Md.; CARPENTER, WM H., Columbia College, N. Y.; CLÉDAT, L., Faculté des Lettres, Lyons, France; COHN, ADOLPHE, Harvard University, Mass.; COOK, A. S., Yale University; COSIJN, P. J., University of Leyden, Holland; CRANE, T. F., Cornell University, N. Y.; DAVIDSON, THOMAS, Orange, N. J.; EGGE, ALBERT E., St. Olaf's College, Minn.; FAY, E. A., National Deaf-Mute College, Washington, D. C.; FORTIER, ALCÉE, Tulane University, La.; GARNER, SAMUEL, U. S. Naval Academy; GERBER, A., Earlham College, Ind.; GRANDGENT, CHARLES, Harvard University, Mass.; GUMMERE, F. B., The Swain Free School, Mass.; HART, J. M., University of Cincinnati, Ohio; HEMPL, GEO., University of Michigan; HUSS, H. C. O., Princeton College, N. J.; VON JAGEMANN, H. C. G., Harvard University; KARSTEN, GUSTAF, University of Indiana, Ind.; LANG, HENRY R., The Swain Free School, Mass.; LEARNED, M. D., Johns Hopkins University, Md.; LEYH, EDW. F., Baltimore, Md; LODEMAN, A., State Normal School, Mich.; MORFILL, W. R., Oxford, England; MCCABE, T., Johns Hopkins University, Md.; MCELROY, JOHN G. R., University of Pennsylvania, Pa.; O'CONNOR, B. F., Columbia College, N. Y.; PRIMER, SYLVESTER, Providence, R. I.; SCHELE DE VERE, M., University of Virginia, Va.; SCHILLING, HUGO, Wittenberg College, Ohio; SHELDON, EDW. S., Harvard University, Mass.; SHEPHERD, H. E., College of Charleston, S. C.; SCHMIDT, H., University of Deseret, Salt Lake City, Utah; SIEVERS, EDUARD, University of Tübingen, Germany; SMYTH, A. H., High School of Philadelphia, Pa.; STODDARD, FRANCIS H., University of City of New York; STÜRZINGER, J. J., Bryn Mawr College, Pa.; THOMAS, CALVIN, University of Michigan, Mich.; WALTER, E. L., University of Michigan, Mich.; WARREN, F. M., Johns Hopkins University, Md.; WHITE, H. S., Cornell University, N. Y.

SUBSCRIPTION PRICE ONE DOLLAR AND FIFTY CENTS PER ANNUM,
Payable in Advance.
FOREIGN COUNTRIES $1.75 PER ANNUM.
SINGLE COPIES TWENTY CENTS.
Specimen pages sent on application.

Subscriptions, advertisements and all business communications should be addressed to the

MANAGING EDITOR OF MODERN LANGUAGE NOTES,
JOHNS HOPKINS UNIVERSITY, BALTIMORE, MD.

STUDIES IN HISTORY, ECONOMICS AND PUBLIC LAW.

EDITED BY

THE UNIVERSITY FACULTY OF POLITICAL SCIENCE OF COLUMBIA COLLEGE.

The monographs are chosen mainly from among the doctors' dissertations in Political Science, but are not necessarily confined to these. Only those studies are included which form a distinct contribution to science and which are positive works of original research. The monographs are published at irregular intervals, but are paged consecutively as well as separately, so as to form completed volumes.

The first four numbers in the series are:

1. **The Divorce Problem—A Study in Statistics.** By WALTER F. WILLCOX, Ph. D. Price 50 cents.

2. **The History of Tariff Administration in the United States, from Colonial Times to the McKinley Administrative Bill.** By JOHN DEAN GOSS, Ph. D. Price 50 cents.

3. **History of Municipal Land Ownership on Manhattan Island.** By GEORGE ASHTON BLACK, Ph. D. (In Press.)

4. **Financial History of Massachusetts.** By CHARLES H. J. DOUGLAS. (*Ready in October, 1891.*)

Other numbers will be announced hereafter.

For further particulars apply to

PROFESSOR EDWIN R. A. SELIGMAN,
Columbia College, New York.

American Economic Association.

PUBLICATIONS.

A series of monographs on a great variety of economic subjects, treated in a scientific manner by authors well known in the line of work they here represent.

Among the subjects presented are Coöperation, Socialism, the Laboring Classes, Wages, Capital, Money, Finance, Statistics, Prices, the Relation of the State and Municipality to Private Industry and various Public Works, the Railway Question, Road Legislation, the English Woolen Industry, and numerous other topics of a like nature.

The latest publication is that for January–March, 1891,—Vol. VI, No. 1-2:—

REPORT OF THE PROCEEDINGS AT THE FOURTH ANNUAL MEETING.

Price One Dollar.

Five volumes of these publications, containing thirty numbers, are now complete.

They will be sent:—bound in Cloth, at $5 each; any two for $9; any three for $13; any four for $17; and all five for $21; all five volumes bound in half morocco, $23.50; single volumes in half morocco $5.50. Unbound, $4 per volume. Forwarded post-paid.
Subscription to Vol. VI, $4.00.

Annual membership $3; life membership $50.

Address,

RICHARD T. ELY, Secretary.

Baltimore, Md.

THE REPUBLIC OF NEW HAVEN.
A History of Municipal Evolution.
By CHARLES H. LEVERMORE, Ph. D.
(*Extra Volume One of Studies in Historical and Political Science.*)
The volume comprises 342 pages octavo, with various diagrams and an index. It will be sold, bound in cloth, at $2.00.

PHILADELPHIA, 1681-1887:
A History of Municipal Development.
By EDWARD P. ALLINSON, A. M., AND BOIES PENROSE, A. B.
(*Extra Volume Two of Studies in Historical and Political Science.*)
The volume comprises 444 pages, octavo, and will be sold, bound in cloth, at $3.00; in law-sheep at $3.50.

Baltimore and the Nineteenth of April, 1861.
A Study of the War.
By GEORGE WILLIAM BROWN,
Chief Judge of the Supreme Bench of Baltimore and Mayor of the City in 1861.
(*Extra Volume Three of Studies in Historical and Political Science.*)
The volume comprises 176 pages, octavo, and will be sold, bound in cloth, at $1.

Local Constitutional History of the United States.
By GEORGE E. HOWARD,
Professor of History in the University of Nebraska.
(*Extra Volumes Four and Five of Studies in Historical and Political Science.*)

Volume I.—Development of the Township, Hundred, and Shire, is now ready. 542 pp. 8vo. Cloth, Price, $3.00.
Volume II.—Development of the City and Local Magistracies. In preparation.

THE NEGRO IN MARYLAND:
A STUDY OF THE INSTITUTION OF SLAVERY.
By JEFFREY R. BRACKETT, Ph. D.
(*Extra Volume Six of Studies in Historical and Political Science.*)

270 pages, octavo, in cloth. $2.00.

The extra volumes are sold at reduced rates to regular subscribers to the "Studies."

NOTES SUPPLEMENTARY TO THE STUDIES.

The publication of a series of *Notes* was begun in January, 1889. The following have thus far been issued:

MUNICIPAL GOVERNMENT IN ENGLAND. By Dr. ALBERT SHAW, of Minneapolis, Reader on Municipal Government, J. H. U.
SOCIAL WORK IN AUSTRALIA AND LONDON. By WILLIAM GREY, of the Denison Club, London.
ENCOURAGEMENT OF HIGHER EDUCATION. By Professor HERBERT B. ADAMS.
THE PROBLEM OF CITY GOVERNMENT. By Hon. SETH LOW, President of Columbia College.
THE LIBRARIES OF BALTIMORE. By Mr. P. R. UHLER, of the Peabody Institute.
WORK AMONG THE WORKINGWOMEN IN BALTIMORE. By Professor H. B. ADAMS.
CHARITIES: THE RELATION OF THE STATE, THE CITY, AND THE INDIVIDUAL TO MODERN PHILANTHROPIC WORK. By A. G. WARNER, Ph. D., sometime General Secretary of the Charity Organization Society of Baltimore, now Associate Professor in the University of Nebraska.
LAW AND HISTORY. By WALTER B. SCAIFE, LL. B., Ph. D. (Vienna), Reader on Historical Geography in the Johns Hopkins University.
THE NEEDS OF SELF-SUPPORTING WOMEN. By Miss CLARE DE GRAFFENREID, of the Department of Labor, Washington, D. C.
THE ENOCH PRATT FREE LIBRARY. By LEWIS H. STEINER, Litt. D.
EARLY PRESBYTERIANISM IN MARYLAND. By Rev. J. W. MCILVAIN.
THE EDUCATIONAL ASPECT OF THE U. S. NATIONAL MUSEUM. By Professor O. T. MASON.
UNIVERSITY EXTENSION AND THE UNIVERSITY OF THE FUTURE. By R. G. MOULTON.

These *Notes* are sent without charge to regular subscribers to the Studies. They are sold at five cents each; twenty-five copies will be furnished for $1.00.

ANNUAL SERIES, 1883-1890.

Eight Series of the University Studies are now complete and will be sold, bound in cloth, as follows:

SERIES I.—LOCAL INSTITUTIONS. 479 pp. $4.00.
SERIES II.—INSTITUTIONS AND ECONOMICS. 629 pp. $4.00.
SERIES III.—MARYLAND, VIRGINIA, AND WASHINGTON. 595 pp. $4.00.
SERIES IV.—MUNICIPAL GOVERNMENT AND LAND TENURE. 600 pp. $3.50.
SERIES V.—MUNICIPAL GOVERNMENT, HISTORY AND POLITICS. 559 pp. $3.50.
SERIES VI.—THE HISTORY OF CO-OPERATION IN THE UNITED STATES. 540 pp. $3.50.
SERIES VII.—SOCIAL SCIENCE, MUNICIPAL AND FEDERAL GOVERNMENT. 628 pp. $3.50.
SERIES VIII.—HISTORY, POLITICS, AND EDUCATION. $3.50.

The set of eight volumes is now offered, uniformly bound in cloth, for library use, for $24.00. The eight volumes, with seven extra volumes, "*New Haven*," "*Baltimore*," "*Philadelphia*," "*Local Constitutional History*," *Vol. I*, "*Negro in Maryland*," "*U. S. Supreme Court*," and "*U. S. and Japan*," altogether fifteen volumes in cloth, for $33.00. The seven extra volumes (now ready) will be furnished together for $10.50.

All business communications should be addressed to THE JOHNS HOPKINS PRESS, BALTIMORE, MARYLAND. Subscriptions will also be received, or single copies furnished by any of the following

AMERICAN AGENTS:

New York.—G. P. Putnam's Sons, 27 W. 23d St.
New Haven.—E. P. Judd.
Boston.—Damrell & Upham; W. B. Clarke & Co.
Providence.—Tibbitts & Preston.
Philadelphia.—Porter & Coates; J. B. Lippincott Co.
Washington.—W. H. Lowdermilk & Co.; Brentano's.
Baltimore.—John Murphy & Co.; Cushing & Co.
Cincinnati.—Robert Clarke & Co.
Indianapolis.—Bowen-Merrill Co.
Chicago.—A. C. McClurg & Co.
Louisville.—Flexner & Staadeker.
San Francisco.—Bancroft Company.
New Orleans.—George F. Wharton.
Richmond.—Randolph & English.
Toronto.—Carswell & Co.
Montreal.—William Foster Brown & Co.

EUROPEAN AGENTS:

London.—Kegan Paul, Trench, Trübner & Co.; G. P. Putnam's Sons.
Paris.—A. Hermann, 8 rue de la Sorbonne; Em. Terquem, 31 bis Boulevard Haussmann.
Strassburg.—Karl J. Trübner.
Berlin.—Puttkammer & Mühlbrecht; Mayer & Müller.
Leipzig.—F. A. Brockhaus.
Frankfort.—Joseph Baer & Co.
Turin, Florence, and Rome.—E. Loescher.

NEW EXTRA VOLUMES.

Extra Volume VII, Now Ready.

The Supreme Court of the United States:
Its History and Influence in Our Constitutional System.

BY W. W. WILLOUGHBY,

Fellow in History, Johns Hopkins University.

124 pp. 8vo. Cloth. Price, $1.25.

In this work, published as the seventh extra volume of the Johns Hopkins University Studies in History and Politics, is presented the results of an investigation into the history and development of the Supreme Court of the United States, with a critical examination of its relations with other branches of our federal and state governments.

The subject is treated under the following divisions.

- I. The Judiciaries in the Colonies, and under the Confederation.
- II. The Judiciary in the Convention.
- III. The Judiciary in the State Convention.
- IV. The Establishment and Jurisdiction of the Federal Courts.
- V. The Supreme Court and Congress.
- VI. The Supreme Court and the State Legislatures and Judiciaries.
- VII. The Supreme Court and the Executive.
- VIII. The Federal Judiciary in Politics.
- IX. The Present Condition and Needs of the Supreme Court.
- X. Conclusion. *Appendix:* Key to Reports; Table of Cases; Index.

Extra Volume VIII, Now Ready.

The Intercourse between the United States and Japan.

BY INAZO (OTA) NITOBE,

Associate Professor, Sapporo, Japan.

198 pp. 8vo. Cloth. Price $1.25.

This monograph is a convenient source of reference for information concerning the gradual development of the foreign relations of Japan.

A few of the points dwelt upon are the reasons why Japan closed her ports for two centuries and a half; the means by which Perry opened the country to foreign commerce; and the subsequent manner in which the United States dealt with the Island Empire.

Dr. Nitobe gives a Japanese view of events and of American diplomatic methods. He describes the part our government and its citizens have taken in helping Japan to achieve the rapid changes. We see from a native standpoint the effect of the educational influences in Japan. We discover the causes which have operated to make the large commercial expectations of the United States an unrealized dream.

To the student of History and Politics the book is valuable, not only because of the events it narrates and the statistics it contains, but also because of the justice done the actors in our international history; while to those who are interested in the educational and religious phases of our foreign relations, there is an impartial and clear statement of facts.

Address orders to

THE JOHNS HOPKINS PRESS,

BALTIMORE, MD.

www.ingramcontent.com/pod-product-compliance
Lightning Source LLC
Chambersburg PA
CBHW021944160426
43195CB00011B/1215